EARTHWATCH
The Climate from Space

EARTHWATCH
The Climate
from Space

JOHN E. HARRIES
Associate Director, Space Science Department
Rutherford Appleton Laboratory, Chilton, Didcot, Oxon

ELLIS HORWOOD
NEW YORK   LONDON   TORONTO   SYDNEY   TOKYO   SINGAPORE

First published in 1990 by
ELLIS HORWOOD LIMITED
Market Cross House, Cooper Street,
Chichester, West Sussex, PO19 1EB, England

A division of
Simon & Schuster International Group
A Paramount Communications Company

Typeset by Ellis Horwood Limited
Printed and bound in Great Britain
by Bookcraft (Bath) Limited, Midsomer Norton, Avon

British Library Cataloguing in Publication Data

Earthwatch.
1. Climatology. Use of observation from space vehicles
I. Harries, John E.
551.6
ISBN 0–13–223843–8

Library of Congress Cataloging-in-Publication Data

Earthwatch: the climate from space / editor, John E. Harries.
p. cm. — (Ellis Horwood series in atmospheric science)
Includes biobliographical references and index.
ISBN 0–13–223843–8
1. Climatology. 2. Astronautics in meteorology. I. Harries, John E., 1946– . II. Series.
QC981.E24   1990
551.6–dc20

90–43738
CIP

Table of contents

Foreword

Views of the Earth from space are commonplace; we see them every day on our television screens. Instantly we can see how the clouds are developing and moving over different parts of the globe. Such views help us to realize that we are one world. The circulations of the atmosphere and of the oceans are global in scale; activity in one part of the globe rapidly influences what happens elsewhere. Pollution generated in one part rapidly affects other parts too.

As the human population expands and the demand for resources of all kinds increases, we are all more aware of the vulnerability of our world and its resources to change. Extremes of climate affect all parts of the world at different times as they always have. Now they often have more impact and because of modern communication we are more aware of their existence.

Because of these factors, interest and concern regarding the possibility of climate change and the influence that human activities might have on the climate have increased enormously during the last few years. Demands have been placed on scientists to describe and predict how climate might change next century due to human activities and what impact such changes might have on ecosystems and human communities.

The climate system is, however, extremely complex. Before accurate predictions on a regional scale can be achieved, we have to understand a great deal about the different components of the climate system and how they interact with each other. Observations from space are absolutely key to progress in scientific understanding. Their interpretation is often difficult and complex. This book — *Earthwatch: the climate from space* — is therefore very welcome and timely.

John Harries has been involved in atmospheric observations for many years. When at the National Physical Laboratory he led a small but effective team making spectroscopic measurements of atmospheric composition. For the last ten years or so he has had broader responsibilities at the Rutherford Appleton Laboratory where he is currently the Associate Director in charge of space science. Despite his substantial administration load he still finds time for some scientific work of his own in the field of Earth observation. He has written this volume from the viewpoint of the scientific practitioner and his enthusiasm for the subject comes through. Students and others who read his exposition will not only find the book a source of information but also will pick up some of his catching enthusiasm.

John Houghton

Author's preface

This book is intended as an introductory text for those interested in learning something about our present understanding of the scientific principles lying behind the workings of the Earth's climate system, and how we can use modern space techniques to observe and study the climate, and so to develop that understanding further. This is an exciting, and relatively new, area of scientific endeavour, which calls for imagination and boldness in our scientists: perhaps after reading this book you will feel like playing your own part in what is not just a fascinating area of science, but one of vital importance to the future of not only today's younger generation, but many generations to come.

In writing the book, I have sought to bring together two separate features that are not often linked, but which scientifically are closely connected: on the other hand, a simple description of the basic scientific principles which lie behind the immense complexity of the Earth's climate, and on the other, a review of how we can use the techniques of remote sensing from space to study these scientific principles. From the point of view of the physical scientist concerned with climate research, both these aspects are of interest, and are intimately connected: if, as experimentalists, we are to make measurements that will improve our understanding of the processes, we must have some comprehension of what those processes are; in reverse, if, as theoreticians, we would like to propose measurements to test our theories of climate, we should have a grasp of what measurements are possible and how they are made. Too often, the practitioners of theory and experiment do not talk very much to one another. This book will, I hope, provide an introduction to the subject for new-comers to the field who want something of an overview of how theory and measurements from space come together.

It might help the reader to know that the level of the book was at least intended to be aimed at students at the graduate, undergraduate and school sixth form levels. Whether or not it succeeds remains to be discovered. I have assumed a basic scientific and mathematical literacy, although I have tried to keep the text accessible also to students with a non-scientific background. It is also intended as a useful introductory text and status review for informed laymen, for professionals from other disciplines, and for non-specialist scientists, engineers, administrators and others who may be involved or interested in space programmes or environmental projects.

The book begins with a short introduction, followed by a survey of the scientific principles of the climate system, including its major components: the atmosphere, the oceans, the cryosphere and the land. In this review of the underlying scientific principles, I have attempted to provide a description of most aspects which are of importance, for example the dynamics, the chemistry and the radiation. I make no apologies, however, for emphasizing the radiative aspects: not only do my own interests lean in this direction, but many climate problems (for example the greenhouse effect) are strongly driven by radiation processes. More importantly, however, it is the observation of the radiative fluxes leaving the Earth after having interacted with the atmosphere and the surface that forms the basis of remote sensing from space.

After this scientific review, a more detailed development is then given of three specific climate topics that are often in the news, but which perhaps are not well understood by many people: these are the ozone hole, the greenhouse effect and the 'El Nino': many other examples could have been cited, of course, but this selection provides something of a cross-section of process types, and also illustrates both anthropogenically induced and natural climate fluctuations. Following this, we turn to a consideration of the practical aspects of space observations, including some details of instruments, spacecraft and their orbits. Finally, we examine up-to-date examples of global data that have been obtained from satellites, and look towards future prospects for monitoring the climate from space. While the choice of examples will reflect, of course, this author's own interests in atmospheric science and space observations, the range of material covering also ocean, land and cryospheric aspects will, it is hoped, provide a through introduction to wider reading elsewhere.

This volume has taken some time to produce since its original conception in 1984, owing entirely to the lethargy of the author. In that time, however, we have seen the topic of climate itself change from a little-mentioned subject reserved for specialist conferences, to a highly public matter for debate. This change has been welcome, especially on the part of the politicians who previously had given quite inadequate attention to the threat of climate change. Now there is a recognition in the UK that climate change is important, and, moreover, that space techniques have a vital role to play in the long-term process of understanding how the climate works, and in monitoring how it is changing.

I hope that this book, aimed very purposefully at the student, and the non-expert anxious to gain a better understanding of the climate system as seen from space, will contribute to the important task of education: of creating interest in climate and space in the minds of the public, and in attracting young people to take up the subject as a career. Whether this is as a mathematician, a physicist, an engineer, a space technologist, a chemist, a physical geographer, or any one of a number of other specialists does not matter; climate research is a highly multi-disciplinary subject, and many types of expertise are needed.

Finally, I wish to thank a number of people. First, my wife Sheila and our children Paul, James and Rebecca, for their long sufference of the neglect that writing a book brings; my mother and father for their life-long support; also, my secretary, Kay Knight, and many other colleagues at RAL, including Linda Roberts, Nicky

Smithers and others, for their help; valuable comments were made by Dr John Houghton, who has also kindly contributed a Foreword, and Dr Chris Rapley; last but not least, thanks are due to Felicity Horwood, Karen Rose and the whole team at Ellis Horwood, for their patience and advice.

Acknowledgements

The following acknowledgements are made for permission to publish plates and figures in the text:

Plates 3, 7, 8, 14, 15, 16, 17 and 19 and Figs 1.2, 2.2, 2.23, 3.1, 3.8, 4.12, 4.13, 4.16, 4.18, 4.19 and 5.3 are taken from published NASA documents and are published with the permission of the National Aeronautics and Space Administration, Washington, DC , USA; Figs 2.8, 5.5 and 5.6 were kindly supplied by Dr C. B. Farmer of the Jet Propulsion Laboratory, Pasadena, California, and are published with the permission of NASA, and of the Springer-Verlag publishing company of Heidelberg; Plate 13 was kindly supplied by Dr M. Chahine of the Jet Propolusion Laboratory, and is published with the permission of NASA; Plates 9, 10 and 11 were kindly supplied by Dr J. M. Russell III, of NASA's Langley Research Center, Hampton, Virginia, and are published with the permission of NASA;

Plates 1, 2, 5 and 6 and Fig. 4.20 are published with the permission of the European Space Agency;

Figs 2.6, 2.15, 2.16, 4.14, 4.15 and 4.17, are taken from *Methods of Satellite Oceanography*, by Robert Stewart, and are published by permission of the University of California Press as its publisher;

Fig. 2.1 is published by permission of the World Meterological Organization, who hold the copyright: the figure was adapted from an original figure in the Report of the Panel of Climatic Variation to the US GARP Committee;

Figs 2.4, 2.24 and 2.29 have been taken from Ulaby *et al.*, *Microwave Remote Sensing: Active and Passive,* Vol III, 1986, and are published with permission of Artech House Inc., USA;

Figs 2.7, 3.14, 3.15 and 3.16 have been taken from Manabe (1983) and Saltzmann (1983), and are published with permission of Academic Press;

Fig. 2.14 is taken from M. I. Budyko *Climate and Life,* and is published by permission of Academic Press;

Plate 4 is published by permission of Dornier, prime contractors for the European Space Agency's ERS-1 spacecraft;

Figs 2.17, 2.18, 2.19, 2.26, 2.27, 3.10 and 5.1, are taken from Houghton (Ed.), *The Global Climate,* and are published by permission of the Cambridge University Press;

Figs 3.2, 3.3, 3.4, 3.5, 3.6, 3.7, 3.12 and 3.13 are taken from *Stratospheric Ozone 1987* and *Stratospheric Ozone 1988,* and are published by permission of the Controller of Her Majesty' Stationery Office;

Figs 2.20, 2.21 and 2.22, are taken from *The International Satellite Land Surface Climatology Project Report ISLSCP No. 10. 1988* and are published with permission of the authors, the American Society of Photogrammetry, and the American Institute of Biological Science;

Figs 2.10, 2.11 and 2.13, are taken from *Introduction to Physical Oceanography* by J. A. Knauss, and are published by permission of Prentice-Hall Inc., Englewood Cliffs, New Jersey;

Fig. 5.2 is taken from Guymer, *Philosophical Transactions of the Royal Society,* **A309,** 399–414, 1983, and is published with the permission of the Royal Society and the author;

Plate 12 is published with the permission of the Deutsche Forschungsanstalt fur Luft- und Raumfahrt e.V.;

Fig. 2.12 is reprinted from *Understanding Climatic Change* and is published with permission from the National Academy of Sciences, National Academy Press, Washington DC;

Figs 3.12 and 3.13 are based on figures appearing in Ramanathan *et al., J. Geophys. Res.,* **90,** 5561 (1985); Figs 4.6 and 4.7 are taken from Gille and Russell, *J. Geophys. Res.,* **89,** 5129 and 5130 (1984); copyright by the American Geophysical Union, and published with their permission;

Fig 2.28 is taken from Rott and Soegaard, *Zeitschrift fur Gletscherkunde und Glazialgeologie,* **23,** 115–121 (1987), and is published with permission of the Universitat Wagner, Innsbruck, Austria;

Figs 3.17 and 3.18, are taken from Bjerknes, *Monthly Weather Review,* **97,** 163–172; Fig. 5.4 is taken from Ardanuy *et al., J. Climate,* **22,** 766–799 (1989), they are published with the permission of the American Meteorological Society;

Fig. 3.11 is taken from Jones *et al., Nature,* **322,** 430–434 (1986), and Figs 3.19 and

3.20 from Gill and Rasmussen, *Nature,* **306**, 229–234 (1983); they are reprinted by permission from Macmillan Magazines Ltd, Copyright 1986 and 1983;

Fig. 3.9 is taken from A. S. Monin, *An Introduction to the theory of Climate,* and is published with permission of D. Reidel Publishing Co.;

Fig. 2.25 is taken from Choudhury, *International J. Remote Sensing,* **10**, 1579–1605 (1989); and Plate 18 from Justice *et al., ibid.,* **10**, 1607–1632 (1989); they are published with the permission of Taylor and Francis Ltd and the authors;

Figs 4.8, 4.9 and 4.10 are taken from Waters, *Atmos. Res.* **23**, 391–410 (1989), and are published with permission of Elsevier Science Publishers and the author;

Fig. 4.11 is taken from the Microwave Limb Sounder proposal to the European Space Agency, by J. W. Waters, G. E. Peckham *et al.*;

Fig. 1.1 is published with permission of the Royal Geographical Society.

List of illustrations

List of tables

1

Introduction

1.1 BACKGROUND TO THE BOOK

The aim of this book is to bring together in a single, introductory work a report on our scientific understanding of some of the critical environmental issues which threaten our global climate system, and a description of the new space systems and measurement techniques which we can now marshal to address these problems. In a way, the book is a status report, because developments are occurring so rapidly, in a fascinating area of man's endeavour which is sure to grow in importance in coming years. The use of spacecraft to study our planet on a global scale is, of course, a rather recent new development in mankind's ability to observe and understand the nature of the world in which we live. We can now employ both manned and unmanned space missions for this purpose, using sophisticated automatic cameras, spectrometers and other instruments to probe the secrets of our atmosphere and climate system.

From space we have gained a quite new perspective of our planet, its oceans, atmosphere, polar regions and land surface. This new perspective, which is captured in the striking photographs of the Earth now available from spacecraft like Meteosat (see Plate 1), is one of a beautiful, unique (at least locally in the universe) but perhaps fragile planet — the home of mankind. We can see more clearly how insignificant we are on the scale of stars and galaxies, which perhaps teaches us some humility. We can also appreciate more easily how ridiculous it is to spend so much of our time and our resources in fighting and killing one another and in destroying our cities, homes and environment, and in damaging the very products of our own creative efforts. Many of us find that this new view of our planet strengthens our belief in an over-all benevolent and positive force behind creation. Whatever our beliefs, however, the new perspective is striking, it is impressive, it is humbling and inspiring; it is a new perspective which, like the nuclear bomb, forces us to change our views of the world.

The advantages that we gain through the use of space need to be considerable, of course, in order to justify the high costs of space missions, and indeed they are. Space gives us a quite unique global view of the Earth, with a comprehensive coverage and

sampling capability that would be quite impossible if we were constrained to make all
our observations from the Earth's surface or even from aircraft. While it might be
possible to monitor large parts of the northern hemisphere without the use of space,
this would be quite impossible to do regularly and comprehensively for much of the
planet, especially the Pacific and southern oceans, the poles and other remote or
inhospitable regions. From space we can observe all parts of the globe regularly,
using small numbers of well-calibrated instruments, all of which contributes to
reducing errors of observation. It is important not to ignore the problems that also
are associated with space techniques, of course. Thus, careful attention has to be
paid to unscrambling the variations of observed processes which are due to fluctuations in time from those variations due to changes in space, because, of course, one
space instrument cannot observe all the Earth continuously. Also, achieving
extremely high spatial and temporal resolution of local processes is difficult from
space, and often is better done using more local measurement techniques. Nevertheless, space observations are an indispensable component in our assembly of techniques for studying the global environment, indeed a component which is likely to
grow in importance in the future.

The new perspective we gain from space also gives us a clearer view of our
responsibilities: there, before us, is a planet for which we are responsible, for which
we are the custodians. Our growing technological power not only gives us this new
perspective, but gives us also the capability of damaging the planet. The atmosphere,
the oceans, the cryosphere and the land are all fairly robust systems that have
developed over millions of years: but there is a limit to the damage we can inflict upon
them, as is now clear. The pressure of mankind's increasing numbers is now
becoming too great to ignore.

In recent years and months most people will have read in their newspapers, and
seen and heard on TV and radio, of more and more existing and potential threats to
our planet: floods, droughts, climate change, holes in the ozone layer, greenhouse
warming of the planet, acid rain and so on. This is not (just) a new fad of the news
media, but a recognition that we maybe need to think and act more carefully in how
we treat our planet.

Thus, scientists have now taken up the challenge of understanding better the way
in which the systems on planet Earth work, and through that understanding learning
how better to protect and look after it. We shall see that observations of the Earth
from space have become a vitally important part of the process of reaching that better
understanding. On the international front, this has been recognized in major
cooperative research programmes to study the Earth's atmosphere and climate, such
as the Global Atmospheric Research Programme (GARP), the World Climate
Research Programme (WCRP), the international weather-monitoring system
known as the World Weather Watch, and in future the International Geosphere–
Biosphere Programme (IGBP), all of which include substantial elements devoted to
global observations from space. It is, of course, important to keep a sense of
perspective and recognize that techniques other than those using space also have an
important role to play in studying and monitoring our planet: nevertheless, the
global perspective obtained from space is unique.

Against this background, this book is an introductory text which tries to explain
to the student and the non-specialist something about the physical basis underlying

some important processes going on within the climate system, and how the new techniques of space observations are being used to study these processes. I hope that the book will give the reader a taste of the fascination of the study of our planet from space, and perhaps some insight into the sheer beauty of the Earth that is revealed for us.

1.2 HISTORICAL PERSPECTIVES OF THE PLANET EARTH

It is interesting to illustrate the growth in man's awareness of his planet as it is linked to his ability to observe greater and greater parts of it. Throughout history our

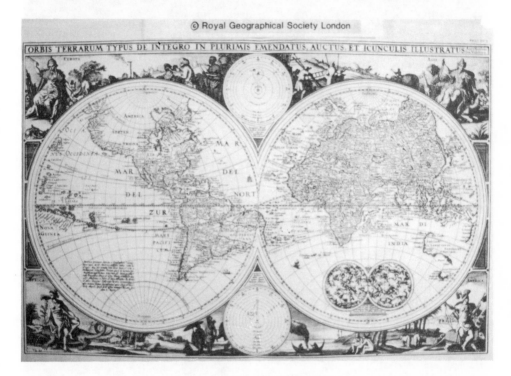

Fig. 1.1 — A map of the Earth in 1657.

perception of the scale and nature of the Earth has grown with our technological ability to study and survey it on ever greater scales. Ancient civilizations depended on memory to record their view of a small part of the planet, the known part, and mainly on folklore and imagination to account for what lay beyond. Technological developments such as writing, mapmaking, navigation, sea travel, each led to an increase in the accuracy of records, and in the scale and extent of observations that provided the basis for our knowledge of the world. However, even by the early middle ages man's perceptions of the planet Earth were still very limited, and all sorts of folk tales were invoked to explain what happened beyond the known limits. Fig. 1.1 shows a map of the world dating from 1657. The observations that formed the basis of this map were simple geometric ones, using the principles of triangulation

Fig. 1.2 — Earthrise seen from the Moon during the Apollo programme.

and navigation, and the maps were, of course, developed primarily because of a driving economic need — trade. It is true that some of the most spectacular feats of exploration and observation were undertaken purely to satisfy the human 'wander-lust' and the desire to seek out the unknown: but underlying these pinnacles of achievement was a somewhat less noble, yet extremely powerful, motive force: the hope of making a fortune through trade and commerce abroad. We must not take too high-and-mighty a view of this, however, because often it is true that fairly graceless human drives can lead to supreme achievements: indeed, we might argue that the conquest of space today might not be progressing as fast were there not huge profits to be made from industrial and particularly military contracts; nevertheless, high cultural achievements benefiting all mankind do still emerge, as we shall see.

Returning to Fig. 1.1, we can see the boundaries of the then-known world, by comparison with a modern map of the globe. Much of what was drawn in maps such as these was, indeed, mere speculation. Adding to this picture of the world the fact that man had but a poor understanding of why the atmosphere behaved as it did, why earthquakes and storms occurred, and why volcanoes erupted, then we can under-stand that without the safety net of religion, or in some cases superstition, the world would have been a very scary and unpredictable place indeed.

Re-reading the last sentence, you may feel that not much has changed. However, what has indisputably changed is the position of the boundary between what we actually know about our planet and the region of speculation. Now, as Fig. 1.2 illustrates most powerfully, we have the capability of using spaceflight to look at our planet from great distances, and to see its unity, its beauty and possibly its fragility. We can probe the depths of the atmosphere, the vast empty ocean regions, the zones of ice and snow, and we can map the whole surface in great detail. From this information we can now take a global view of our finite planet. This global view has given us the capability to think more clearly about the vastly complex system of balancing processes that form the environment that we know, the balance between heat and cold, wind and calm, sea and air and indeed, flood and drought. This balance, this equilibrium between a bewildering array of different cycles and processes is our climate. What is climate? How stable is it? What are the most sensitive triggers to changing climate? How can we use space flight to study the climate globally? What new, advanced space instruments do we have available to do this? How do we interpret the data from such sensors? For a closer look into some of these and other questions, now read on.

2

The climate system

What is 'climate'? It is a thing which we all feel we understand, but which proves difficult to define in any detail. It can be thought of as an average state of the weather, which smoothes out the short-period fluctuations but preserves the longer-term significant trends. The recently published report of the Inter-Government Panel on Climate Change has pointed out that the description of the climate is based on the average state of various components of the weather, together with the variability of those components, the averages being taken over periods ranging from years to centuries. The 'weather' in this sense includes, of course, not only the state of the atmosphere, but that also of the oceans, cryosphere and so on.

Climate also has a more subjective meaning, being the perceived influence of the state of the atmosphere on our daily lives on the surface of the Earth. Scientists have attempted more complicated definitions in order to provide an adequate basis for a mathematical description of the climate system, but for our present purposes we can make do with the more subjective definition just given, taking note of two key characteristics of the climate which will recur as themes throughout this book:

- The climate system is an extremely complex and inter-connected (coupled) system;
- The ultimate predictability of the climate is not well understood.

2.1 BASIC PRINCIPLES

2.1.1 Energy

The energy necessary to provide the basic power source to our planet comes almost entirely from the Sun. Other sources are by comparison insignificant (e.g. natural nuclear fission, moonlight, starlight). The core of the Earth is hot, and is cooling on a time scale that is irrelevant to all but the longest geological time scales. Later in this book we will see that the energy from the sun arrives at the Earth in the form of electromagnetic radiation over a wide range of wavelengths. The peak of this incoming radiation occurs in the so-called visible region, at wavelengths of between

0.4 and 0.7 microns or so, and it is of course no accident that our eyes have evolved to be sensitive in this waveband, for maximum efficiency. It is also important to note that it is only because the atmosphere is highly transparent (in the absence of cloud) at visible wavelengths that we have evolved in this way. At other wavelengths which carry significant amounts of incoming solar radiation, e.g. parts of the infrared or the ultraviolet, the atmosphere is quite heavily absorbing, and little radiation reaches the ground. We shall develop some of the physics behind these processes later on.

2.1.2 Complexity of processes

As we have seen, this thing called 'climate' is a somewhat intangible entity. It comprises the net balance of a large number of physical and chemical processes, involving the oceans, the air, clouds, ice and snow, the land, vegetation and indeed living creatures. It encompasses physical processes like freezing, melting, winds and ocean currents, and also chemical and radiative processes like the production of methane from swamps, and of pollutants by man, and the absorption and scattering of solar radiation by molecules of the air, dust and cloud droplets and particles. In a sense, 'climate' and the complexity of all the physical and chemical processes taking place in our world is analogous to 'health', and the complexity of physical and chemical processes taking place in our bodies. In each case it is the net balance, the state of well-being, of an unimaginably complex system which is of concern.

Each different component of the climate system responds to a change of stimulus at a different rate: thus, if the sun suddenly appears following a long period of cloudiness, the soil of the Earth will heat up rather quickly (in hours) while the ocean will require days to weeks to reach a new equilibrium temperature. This is simply because the 'heat capacity' of a cubic metre of ocean is much greater than the heat capacity of a corresponding cubic metre of soil. Therefore more heat energy must go into the water than the soil in order to raise the temperature by a given amount. In other words, because the quantity of energy per unit time in the Sun's rays is constant, the temperature of the water takes longer to rise.

This is a rather simple picture, but it does serve to illustrate one of the fundamental difficulties in achieving an understanding of the Earth's climate. It is not sufficient (as it is, say, in weather forecasting — which is already difficult enough!) to consider a restricted range of processes, i.e. those with typical time constants of hours to days. Instead we must understand processes on time scales of minutes, or shorter, to thousands of years. What is even more difficult, we must comprehend how this bewildering array of processes interact with one another.

Fig. 2.1 is a diagram widely used within the climate research community. It illustrates the complexity of interacting processes that make up the climate system. It is an attempt to illustrate what goes into the climate 'brew'. The diagram shows the primary input of radiation from the sun at the top of the atmosphere, balanced by the outgoing terrestrial radiation. Without a very close net balance, of course, the planet would rapidly heat up or cool down. Within the atmosphere, the constituents of the air such as water vapour H_2O, nitrogen N_2, oxygen O_2, carbon dioxide CO_2, ozone O_3, etc., play a role in absorbing and re-emitting radiation in the ultraviolet, the visible and the infrared, while clouds have an important role as reflectors of solar radiation and thermal emitters in their own right. Lower down, interactions between the atmosphere and the oceans, ice and land surface are indicated. These processes

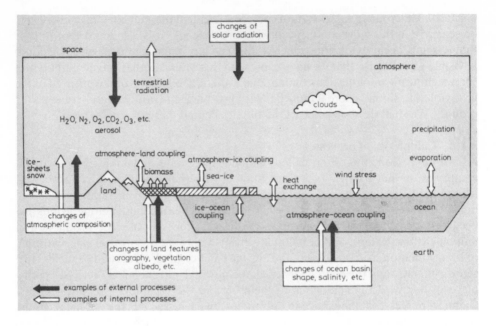

Fig. 2.1 — A schematic diagram of the climate system and its internal interactions.

are many and complicated. The boxes towards the bottom of the diagram illustrate how some changes can be forced in the system, for example by natural or anthropogenically induced changes in atmospheric composition, modification to vegetation, orography, etc.

An important concept to understand is that of 'non-linear feedback' processes in the climate system. We have already laid great stress on the complexity of the climate system, resulting from the many separate and distinct processes which go to make up the climate. The reality is further complicated by the fact that these many processes can interact with each other, or 'feedback' on one another. Sometimes the interaction is simple and direct — in other words, if we change one parameter (for example, the degree of cloudiness) then another parameter will change in proportion (e.g. surface temperature). Sometimes, however, the feedback is 'nonlinear' — in other words, a change in one parameter or process can lead to a much amplified (or diminished) effect elsewhere in the total system. The nonlinear feedback can, of course, be positive or negative (i.e. it reinforces or opposes the original process). It is this non-linearity of feedback processes in the Earth's climate system that makes understanding and predicting the system so difficult, and, it must be said, such a fascinating subject. It is also the basis of our concern that the Earth's climate might not be the stable, well-behaved system we would like to believe.

To illustrate the concept of non-linear feedback, we might consider one or two simple examples. Consider the polar regions of the Earth. Imagine that some influence causes the temperature of the surface layers of the global oceans to rise. This would cause a melting of the marginal ice zones (for instance a northward migration of the sea–ice limits in the northern hemisphere). This in turn would reduce the planetary albedo (reflectivity), with the result that more solar radiation would be absorbed by the Earth's surface, adding to the warming effect. This is positive feedback, and would be non-linear in that, undisturbed, the process could become 'runaway'. However, there would be many other competing processes; and to illustrate one which could produce a negative feedback, let us take our thought experiment further. In our example, the temperature of the oceans rises further owing to the albedo feedback, and so the evaporation of the water increases (a highly non-linear process, as it happens), and so the degree of cloudiness and overall haziness (turbidity) will increase. This would have the effect of increasing the reflection of solar radiation to space, and so lead to a cooling of the system, which would tend to reduce the original warming.

Imagine the difficulty of advising the politician and the planner on the net effect of the original change in this example. It would be a brave person who would say with any certainty what the net balance would be of just a few of these processes, let alone of the real atmosphere with countless other feedback processes to take into account. This is why we need to build theoretical models of the utmost complexity to simulate the climate system. This is also why we need to make global measurements of key parameters from space, in order to test these models and ensure they are accurate, as well as to monitor the state of the climate and keep our eyes open for unpredicted events. We will come back to some of these topics later in this chapter and elsewhere in this book.

2.1.3 Heat capacity

Here it will be useful to introduce some very simple physics. We can imagine five basic components of the climate system: the atmosphere (which is the most immediate indicator of climatic variation); the upper layer of the oceans (say the uppermost 100 m or so); the rest of the ocean depths (which move and respond to changes much more sluggishly than the upper layers); the so-called 'cryosphere' (ice sheets, sea ice, snow); and the land. More complex models of the climate can include some rather obvious subdivisions of these classes (e.g. arid land, forested land; clear air and cloud; etc.) but for our purpose of illustration the present simple model is quite sufficient.

The thermal properties of each of these components can be described by a few very simple parameters:

Temperature	T (units K)
Mass	M (kg)
Volume	V (m^3)
Specific heat	S (J kg^{-1} K^{-1})

Using T, M, V, and S we can develop some rather interesting and useful concepts. First, the quantity of heat, Q (joules), which will cause the temperature of a mass M of material in a unit volume, with a specific heat, S, to rise by T in one second is given by

$$Q = MST .$$

(2.1)

This can be simply arrived at by straightforward proportionality arguments. On the basis of equation (2.1) it is a simple matter to explain why the different components of the atmosphere respond with different time constants, ϕ, to thermal changes. A mathematical differentiation of equation (2.1) for unit volume gives:

$$\frac{dQ}{dt} = \rho S \frac{dT}{dt} ,$$

(2.2)

where ρ = density, or,

$$\frac{dT}{dt} = \frac{1}{\rho S} \frac{dQ}{dt} ,$$

(2.3)

from which we can see that for a unit energy input rate

$$\frac{dQ}{dt} = 1 ,$$

then the time, ϕ, taken for a rise of 1 K in temperature is:

$$\phi = \rho S .$$

(2.4)

This tells us that the oceans, which have large values of ρ and S, respond slowly to thermal changes, and act as stabilizers in the climate system. The atmosphere, on the other hand, can respond relatively quickly to changes in the forcing function. The land surface falls in between the other two cases, in both density and thermal capacity. Thus, as we know from experience, summer–winter temperature differences on land are much greater than in the oceans, with the result that mid-continental climates are more extreme than coastal climates. (We all prefer to spend hot summer days by the oceans, which provide a welcome cooling effect; and winters are colder and more severe typically in continental regions than in maritime regions.) In practice the picture is more complicated than this: the much larger thermal capacity of the oceans is also greatly influenced by the fact that the oceans are able to move, and to transport heat throughout a considerable depth, whereas the land surface is fixed, and heat can be transported only by conduction, a much slower process. Also, the partial transparency of the oceans allows radiation energy to be deposited directly in depth, which is not the case for the land. Nevertheless, this example has illustrated the important fact that any accurate theory of the climate system must be able to accommodate energy exchange and other processes over a wide range of differing time scales, and must accommodate not just the processes that go on in the atmosphere, but those in the oceans and the land, and the exchange processes between these various regions.

2.1.4 Characteristic time and space scales

To model the actual response of the climate system to an external forcing requires that we consider further than just the simple thermal capacity of the main elements, or domains, of the climate system. We also need to know the characteristic time and space scales associated with them, that is, the typical time taken to relax back to an equilibrium state from an external perturbation (e.g. a change in the solar constant), and the typical distance scales over which different climatic processes occur.

Saltzman (1983) has presented a valuable study of the relaxation times of the principle component domains of the climate. He defines a thermal relaxation time, ε_T, (which he terms an equilibration time), which is similar to our ϕ, and which for a geophysical fluid includes terms to account for

— radiative cooling rate, b,
— direct heat transfer rate (conduction), c,
— dynamical heat transfer rate (through eddy motions in the vertical and horizontal), k_v, k_h.

$$\varepsilon_T = \frac{1}{(b + c + k_v + k_h)}. \tag{2.5}$$

Table 2.1 indicates some of the parameters that entered into Saltzman's calculations, including his assessment of characteristic horizontal and vertical length scales. He demonstrated that if two climatic domains (or regions) have similar relaxation times, then a climate model must solve the equations governing processes in each domain simultaneously, since the processes can interact continuously. However, those domains with large values of ε_T will tend to 'carry along' the domains with smaller values of ε_T, which because of their faster relaxation times tend to adjust rapidly to the changing boundary conditions imposed by the slower response domains. Saltzman's analysis indicates that, though the deep ocean and the large ice masses can be considered essentially independently of the atmosphere, (in Table 2.1 we can see that these domains have very large values of ρS or ε_T), this is not so for the surface layers of the oceans or the land. Therefore, accurate climate models must include these faster-response climatic domains.

A separate study by NASA (NASA, 1988a) has also addressed the question of time and space scales associated with a number of climatic processes. Fig. 2.2, taken from that study, shows on a logarithmic scale the characteristic time and space scales of processes ranging from very long-term, large-scale processes like planetary formation and plate tectonics, to short-term, localized processes such as volcanic eruptions and atmospheric turbulence. In between these extremes are processes which we recognize as being of importance to the trends in 'climate', such as CO_2 variations, deep ocean motions, glaciation, and so on.

2.1.5 Radiation balance

If the Earth had no atmosphere, we could write down a very simple equation to tell us what the mean surface equilibrium temperature would be, simply by balancing the incoming and outgoing radiation, on the assumption that the Earth is in a state of near-equilibrium. If F is the flux of solar radiation at the distance, R, of the Earth from the Sun, a is the fractional planetary albedo (or reflectivity, averaged over all wavelengths) r is the Earth's radius, and T_e is the equilibrium temperature of the planet, then we know from the Stefan–Boltzman law (Goody, 1964) that the emitted radiation energy is

Table 2.1 — Parameters describing various climate domains (from Saltzmann, 1983)

Climate domain	Area, A (10^12 m^2)	Depth, D (m)	Density (kg/m^3)	Mass, M (10^18 kg)	Specific heat, S (10^3 J/kg K)	Thermal mass, MS (10^21 J/K)	Typical horizontal scale, L (m)	Relaxation time, T (s)
Atmosphere								
Free	510	10 000	0–1	5	1	5	10^6	10^6–10^7
Boundary layer	510	1000	1	0.5	1	0.5	10^6	10^5
Ocean								
Mixed layer	334	100	1000	34	4	100	10^5	10^6–10^7
Deep	361	4000	1000	1400	4	5000	10^6–10^7	10^{10}–10^{11}
Cryosphere								
Ice sheets	14	1000	1000	10	2	10	—	10^{12}
Glaciers	1	100	1000	0.1	2	0.1	—	10^{10}
Snow/ice	80	1	500	0.1	2	0.1	—	10^5
Sea ice	30	1–100	1000	0.5	2	0.1–1	—	10^6–10^{10}
Land								
Lithosphere	131	2	3000	1	0.8	1	—	10^6
Lakes/rivers	2	100	1000	0.2	4	1	—	10^6

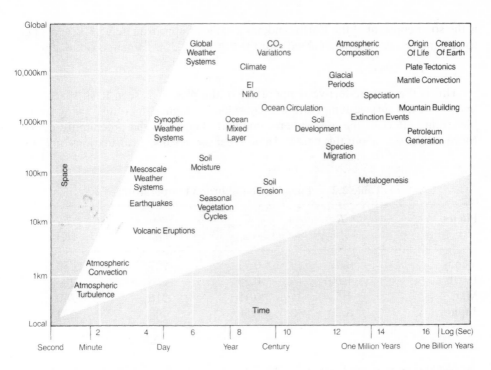

Fig. 2.2 — Illustrating the range of characteristic time and space scales of processes in the climate system.

$$E_e = 4\pi r^2 \sigma T_e^4 \; , \tag{2.6}$$

where σ is the Stefan–Boltzman constant, and the absorbed radiation from the Sun is

$$E_a = \pi r^2 (1 - a) \frac{F}{R^2} \; . \tag{2.7}$$

If equilibrium is maintained, equations (2.6) and (2.7) must be equal in value and we obtain

$$4\pi r^2 \sigma T_e^4 = \pi r^2 (1 - a) \frac{F}{R^2} \; , \tag{2.8}$$

or

$$T_e^4 = \frac{(1 - a)}{4\sigma} \frac{F}{R^2} \; . \tag{2.9}$$

The term

$$\frac{F}{R^2} = S_0 \quad \text{for } R = 1$$

is the solar constant at the Earth, and has a value of approximately 1370 watts per square metre. Using a mean albedo for the Earth of 0.3, we obtain a value of

$$T_e = 256 \text{ K} .$$

This is the mean effective temperature of the planet, as seen from space. We know that the surface temperature is considerably higher than this, owing to the blanketing effect of the atmosphere, popularly known as the greenhouse effect. More will be said about this below. Table 2.2 shows results of calculations of T_e for

Table 2.2 — Planetary equilibrium temperatures

Planet	R	a	T_e (K)	T_s (K)
Venus	0.72	0.77	227	750
Earth	1.00	0.30	256	280
Mars	1.52	0.15	216	240
Jupiter	5.20	0.58	98	134

several planets in the Solar System. Also shown are the mean surface temperatures, T_s, now known to exist on these planets, which indicate (by comparison with T_e) how important the greenhouse effect is for Venus in particular, and how relatively insignificant it is for Mars, with its rather thin atmosphere.

2.1.6 Spectroscopy

For our simple understanding of the climate system, and also of the spaceborne observing techniques which we will come to later, it is necessary to understand some basic facts about the spectroscopy of the Earth's atmosphere, oceans, ice and land, or in other words the way in which radiation as a function of wavelength interacts with these elements.

The distribution of radiation as a function of wavelength arriving at the Earth from the Sun follows a general shape known as the Planck function, shown in the lower section of Fig. 2.3, at left for a mean solar temperature of 5750 K. Similarly, the outgoing radiation emitted by the warm Earth as it maintains thermal equilibrium with this incoming energy follows the same general shape, though shifted to longer wavelengths and different in absolute magnitude: this is demonstrated by the right lower curve in the figure, corresponding to a mean terrestrial temperature of 245 K. This behaviour is a consequence of the general rule that higher temperatures are associated with higher frequencies of electromagnetic radiation. In Fig. 2.3, the areas of the two lower curves have been drawn equal, to demonstrate the fact that the total quantities of energy entering and leaving the planet must be in balance: this has been done by plotting the specific intensity at a given wavelength, $B(\lambda, T)$, multiplied by the wavelength, λ. It must be recognized, however, that the specific intensity at any wavelength for a 5750 K body is of course much greater than for a 245 K body.

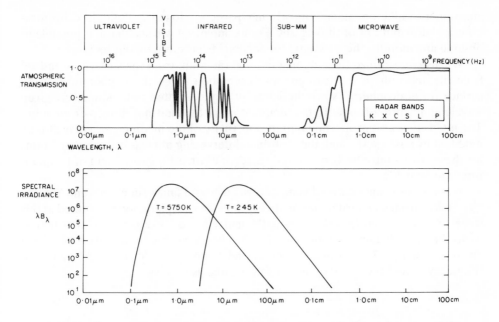

Fig. 2.3 — The transmission spectrum of the atmosphere (upper curve); and (lower curve) the spectral irradiance received at the Earth from the Sun at a mean temperature of 5750 K and the Earth at mean temperature of 245 K (in each case allowing for the solid angle subtended from the Earth).

The equality of energy arises because the 245 K curve extends over a much broader wavelength range, and is distributed over a much greater solid angle and surface area than is the solar beam.

If the Earth's atmosphere were non-absorbing, or absorbed equally strongly at all wavelengths (a 'grey' absorber), then this part of the discussion would not be necessary. This is not the case, however. There are many different molecular species which make up the 'air', and each of these molecules has a characteristic 'signature' absorption spectrum. As a result, the absorption by the atmosphere depends heavily on wavelength, as illustrated in the top part of Fig. 2.3. The consequences of this are many and important, and we must develop some of these here. One of the most important consequences for the present discussion is that this variation of absorption with wavelength gives rise to the greenhouse effect itself: since the atmosphere is quite transparent in the visible region (excluding the effect of clouds), the passage of sunlight to the Earth's surface is fairly unhindered. Consequently, the Earth is heated quite effectively by the incoming visible solar radiation. The Earth re-radiates part of this energy in the infrared, however, where a number of atmospheric molecules (for example CO_2 and H_2O) have very strong absorption bands. Thus, the escape of this thermal radiation to space is hindered, giving rise to a higher surface (and atmospheric) temperature than would occur with a non-absorbing atmosphere.

We will encounter a further consequence of the spectroscopic properties of the atmosphere later. Because the absorption bands are to some extent characteristic signatures of the gases in the atmosphere, we can use measurements of the outgoing atmospheric radiation, made from orbiting spacecraft, to tell us about the concentrations and distributions of those gases. For example, we will see that it is possible to observe and monitor the Antarctic ozone 'hole' from space by this means.

We also may utilize the outgoing thermal radiation from the surface, or indeed from the atmosphere itself, to give us information about the temperature of the surface (or atmosphere), since the intensity of the emitted radiation at any given wavelength is proportional to the temperature of the emitting surface or medium. The intensity is also proportional to the emissivity of the medium, and may also be modified by passage through the absorbing intervening atmosphere, so corrections for these effects must be very carefully made if a correct measurement of temperature is to be made.

We have now encountered some terms, such as 'Planck function', and 'emissivity', taken from standard radiation theory, and some explanation is required at this stage (it is suggested that the non-mathematicians could skip this section if they wish). First, the Planck function describes the energy coming from a 'black body' at a given wavelength λ (or wavenumber v) and temperature T. λ, v, the radiation frequency, f, and the speed of light, c, are related as follows:

$$v = \frac{f}{c}$$

$$\lambda v = 1 \tag{2.10}$$

$$\lambda f = c \ .$$

The expressions for $B(\lambda, T)$ and $B(v, T)$, together with their commonly used units are

$$B(\lambda, T) = \frac{2hc^2\lambda^{-5}}{\exp(hc/k\lambda T) - 1} \, \text{W cm}^{-2}\text{cm}^{-1}\text{ster}^{-1} \ , \tag{2.11}$$

where λ is expressed in units of cm;

$$B(v, T) = \frac{2hv^3/c^2}{\exp(hv/kT) - 1} \quad \text{W cm}^{-2}(\text{cm}^{-1})^{-1}\text{ster}^{-1} \ , \tag{2.12}$$

where v is expressed in cm^{-1}.

The terms emissivity ε_v, reflectivity R_v, and transmissivity τ_v, are related by energy conservation principles as follows

$$\varepsilon_v + R_v + \tau_v = 1 \ . \tag{2.13}$$

In the case of a non-reflecting medium, this reduces to,

$$\varepsilon_v + \tau_v = 1 \ . \tag{2.14}$$

The radiative energy emitted by a real, non-perfect medium is given by

$$E_v = \varepsilon_v B(v, T) \ . \tag{2.15}$$

This expression applies equally, whether we are considering the surface of the land, the oceans, an ice sheet, or indeed a layer of the atmosphere.

Fig. 2.3 shows us that the spectral regions where the absorbing properties of the atmosphere are at a minimum, and which are therefore most useful for remote sensing of the surface, are the visible, between about 0.4 and 0.8 microns, and portions of the infrared, for example in several transparent 'window' regions between about 1 micron and 14 microns. Also shown in the figure is a further region of high 'window' transparency in the submillimetre and microwave wavebands, beyond 1000 microns (or 1 mm) wavelength. Bearing in mind that the Planck curve for thermal emission at temperatures typical of the Earth peaks, as shown in the figure, in the infrared region, we can see that the IR is an optimum choice for remote sensing of the surface and the atmosphere. Though the thermal intensity in the microwave is extremely small, being placed on the long wavelength tail of the Planck curve, nevertheless the microwave region is also usable for remote sensing, on account of the extremely high sensitivity of the heterodyne systems that can be used at microwavelengths, and the high transmission of the atmosphere in this region, which allows a relatively clear view of the surface from space. The broad molecular absorption features of the microwave region are illustrated in Fig. 2.4. The main absorbers are water vapour, H_2O, and oxygen, O_2, and the peak absorptions represent transitions between pure rotational energy states of the molecules.

2.1.7 More advanced radiation transfer

2.1.7.1 Absorption and emission

It is fortunate that, to a very high degree of approximation, the profiles of all molecular absorption lines follow a common profile, determined by the basic physics of the collisional interaction processes between molecules. The fundamental 'line shape' is given by the Lorentz equation, which requires adaptation only at the most detailed level of examination. Based on this fundamental line shape, the strength of absorption as a function of frequency is expressed as the absorption coefficient, α, in the following equation:

$$\alpha_\nu = \frac{S}{\pi} \cdot \frac{\gamma}{(\nu - \nu_0)^2 + \gamma^2} \, , \tag{2.16}$$

where

$$S = \int \alpha_\nu d_\nu \, , \tag{2.17}$$

is the integrated line strength.

The terms in equation (2.16) are defined in Fig. 2.5, which shows the Lorentzian line shape. The wavenumber ν is proportional to the electromagnetic frequency of the absorption line, as we saw earlier in equation (2.10). The wavenumber is inversely proportional to the wavelength, λ,

$$\lambda = \frac{1}{\nu} \, . \tag{2.18}$$

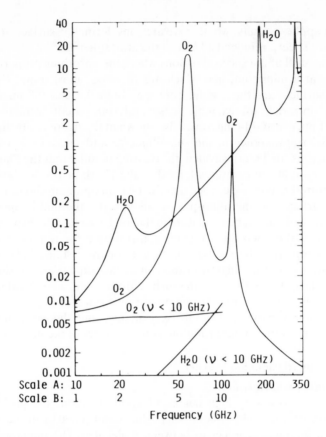

Fig. 2.4 — The microwave spectrum of the Earth's atmosphere, expressed as absorption coefficient in units of dB per km. Separate curves are shown for absorption by water vapour and molecular oxygen.

Here λ would be in units of length (e.g. cm) and ν in units of reciprocal length (e.g. cm^{-1}).

In equation (2.16) γ is the half-width at half-height of the line, and is proportional to the total pressure, P, as in

$$\gamma = \gamma_0 P . \tag{2.19}$$

The general form of equation (2.16) is that of a resonant oscillator, so popular in physics examinations. In this case the 'oscillations' are the vibrations and rotations of the molecules, which can occur at only specific resonant frequencies, determined by quantum mechanics. These equations are particularly relevant to absorption by gases, i.e. to the case of the atmosphere. However, similar equations apply to liquids (the oceans) and solids (i.e. land) except that in these cases, the sharp individual absorption lines in the gaseous case are blurred out due to the stronger interactions between molecules in liquids and solids. There are a number of important appli-

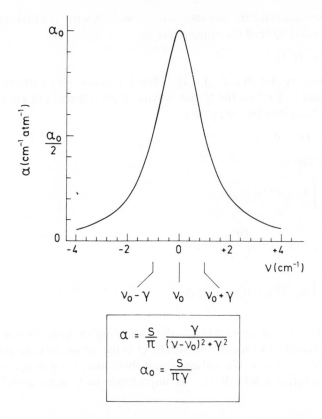

Fig. 2.5 — The Lorentz spectral line shape: α is the absorption coefficient expressed in $cm^{-1} atm^{-1}$; S is the integrated line strength in $cm^{-2} atm^{-1}$; and γ is the half-width of the line at half-height in cm^{-1}.

cations in the remote sensing of atmospheric and other parameters from space which arise from these properties of spectral lines, and which can be exploited in remote sensing experiments, and these will be dealt with later, in Chapter 4.

At any given spectral frequency the transmission of the atmosphere, τ_v, may be written as

$$\tau_v = \exp\left(-\alpha_v \rho z\right) , \tag{2.20}$$

where z in this case denotes path length in the atmosphere, and ρ denotes atmospheric density. Thus, if radiation from the Sun is incident on the atmosphere, the amount to which the incident intensity, $I_v(0)$, is attenuated by the time the radiation reaches a level z in the atmosphere is given by

$$I_v(z) = I_v(0) \exp\left(\int_{top}^{z} \left(-\alpha_v(z')\rho(z')\right) dz'\right) , \tag{2.21}$$

where we have to introduce the integration over path length because both the density, $\rho(z)$, and the absorption coefficient, $\alpha_v(z)$, vary along the path.

Using the expression for the transmission, τ_v, we may write the thermal emission from each elemental layer of the atmosphere as

$$d\varepsilon_v . B(v, T(z)) \ , \tag{2.22}$$

where ε_v is emissivity and $B(v, T_{(z)})$ is the Planck function (see earlier). Since the emissivity is equal to 1 minus the transmissivity, if no reflection of energy occurs, then equation (2.22) may be rewritten as

$$- B(v, T(z)) \, d\tau_v \ , \tag{2.23}$$

and, integrating gives

$$E = \int\int B(v, T(z)) \, d\tau_v \, dv$$

$$= \int\int B(v, T(z)) \left(\frac{\partial \tau_v}{\partial z}\right) dz \, dv$$

$$= \int\int B(v, T(z)) W(v, z) \, dz \, dv \ . \tag{2.24}$$

The term $W(v, z) = \partial\tau/\partial z$ is known as the 'weighting' or 'contribution' function, since it weights the Planck function at each level in the atmosphere to allow for the optical density of each layer. We will encounter this concept of a weighting function in Chapter 4, where we will discuss its importance to remote sounding of the atmosphere.

2.1.7.2 Scattering

Radiation in the atmosphere and oceans is also subject to scattering by molecules and particulates, as well being absorbed according to the rules we have just discussed. In the case of the atmosphere, the particulates include water droplets and ice crystals, fog, clouds, haze, and chemically more complex solid or liquid aerosols, such as produced by volcanoes, smoke, dust and other more complex processes. Moving ahead to Chapter 3 for a moment, we will come across there a more exotic example of aerosol, the liquid/solid low temperature nitric acid/water condensates that are thought to play an important role in the 'ozone hole' problem. In the case of the oceans, particulates due to suspended matter, both organic and inorganic, are strong scatterers.

Scattering is a considerably more difficult subject to deal with in a simple way than is absorption. The magnitude of the scattering efficiency of a particle of radius r depends on the refractive index of the particle relative to the surrounding medium, and therefore on the chemical composition of the particle. It also depends on the wavelength of the radiation being scattered compared to the size of the scatterer.

To develop a simple description of scattering, we have to write down some basic equations which are available from standard text books (e.g. Born and Wolf, 1975; Houghton and Smith, 1966). The refractive index, n, of a material must include an imaginary component, $i\kappa$, if absorption as well as refraction occurs within the material. The complete expression for the so-called 'complex' refractive index is then given by

$$n^* = n - i\kappa .$$ (2.25)

From basic electromagnetic wave theory, the electric field of an e.m. wave can be written as

$$\mathcal{E} = \mathcal{E}_0 \exp(-\omega\kappa z/c)\exp(iw(t - (nz/c)))$$ (2.26)
$$\text{I.} \qquad\qquad \text{II},$$

in which term II represents an oscillating wave, damped by the absorption term I. In equation (2.26), ω is the angular frequency, $\omega = 2\pi f$.

It can be shown (e.g. Houghton and Smith, 1966) that the following relationships exist between absorption coefficient, α, and the imaginary part of n^*; and between the reflectivity R and the real part of n^*, at normal incidence:

$$\alpha = \frac{4\pi\kappa}{\lambda} ,$$ (2.27)

and

$$R = \frac{(n-1)^2 + \kappa^2}{(n+1)^2 + \kappa^2} .$$ (2.28)

For the limiting (and simplified) case where the particles are small compared to the wavelength, the so-called Rayleigh approximation holds, which gives the following expression for the scattering cross-section

$$\sigma_s = \frac{128\pi^5 r^6}{3\lambda^4} \left| \frac{n^2 - 1}{n^2 + 1} \right|^2 .$$ (2.29)

The scattering cross-section, σ_s, is the effective scattering cross-sectional area of the particle compared with its geometric cross-section πr^2. Furthermore, we can envisage the total cross-section of a particle being made up of two components, that due to absorption by the material of the particle itself, and that due to scattering . Thus we can write for the total cross-section

$$\sigma_T = \sigma_s + \sigma_A ,$$ (2.30)

and we can relate σ_T to the spectral intensity, I_T, removed by the particle from a beam of radiation, compared with the incident intensity I_0, as in

$$\frac{\sigma_T}{\pi r^2} = \frac{I_T}{I_0} .$$ (2.31)

Equation (2.29) tells us that the scattering cross-section is strongly dependent on particle size to wavelength ratio (r^6/λ^4), and also on wavelength itself. This is a most important result, and explains why radiation of longer wavelength, e.g. the microwaves, is more suited to applications which involve looking through cloud than radiation of shorter wavelengths, e.g. visible. In a real atmosphere, we experience a somewhat more complex situation than is allowed for in equation (2.29), for example because particles of size approaching and even exceeding the radiation wavelength are encountered, and also because there will inevitably exist a fairly wide range of particle size distributions, both of which essentially lead to other, higher order terms appearing in the equations. Nevertheless, the simple Rayleigh treatment goes a long way to explain the basic phenomena of scattering. For example, we can see why the red sky of sunset occurs: in the long atmospheric paths that exist at sunset, the strong inverse dependence of scattering on wavelength means that blue light from the sun is more strongly scattered out of the beam than red light: thus the light looks red.

Scattering by molecules is much weaker than scattering by particles, except at very short wavelengths, as would be expected from equation (2.29) because of the smaller size of the molecules. The overall broad properties of the scattering atmosphere are summarized in Fig. 2.6 which shows the total 'optical thickness' of a

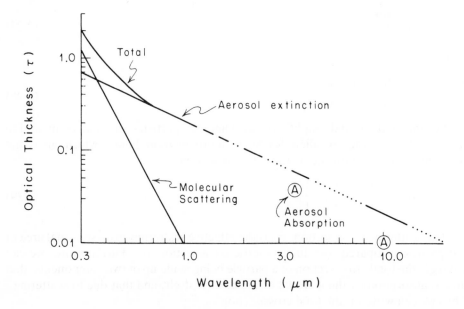

Fig. 2.6 — The optical thickness of the atmosphere due to scattering by aerosols and by molecules, in the visible and infrared spectral regions.

vertical path through a clear atmosphere with a relative humidity of 80%: (optical thickness is proportional to the integral of cross-section over distance along the path, and over all particle sizes which exist along the path). The figure shows that, as expected, aerosol scattering exceeds that due to molecules at all wavelengths longer than about 0.4 microns (the blue end of the visible spectrum).

A short comment on scattering in sea water. As any diver, or armchair watcher of Jacques Cousteau programmes knows, the oceans are very turbid, that is, they are very 'murky' owing to having a high concentration of small suspended particles,

vegetative debris, animal life, etc., and so scatter light strongly. The scattering effect
is so strong that visibility is often limited to tens of metres or less (see Knauss, 1978).
Though difficult to model theoretically, it is evident that scattering has a powerful
influence on the way that visible and infrared radiation are absorbed and distributed
by the various layers of the atmosphere and oceans, and so is an important radiative
process in the balance of the climate.

2.1.8   Variability and predictability of climate

The climate has exhibited natural variability in the past. Evidence from geological
deposits, ice cores, deep-sea cores, tree rings, isotope records and other sources has
clearly shown that this is so. For example, Fig. 2.7 shows a compilation of climatic

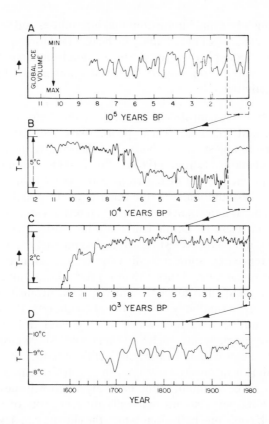

Fig. 2.7 — Selected climatic time series for the past 1 million years: A, from oxygen isotope
variations in deep-sea cores; B and C, from ice cores; and D, from thermometric
measurements.

records covering the past 1 million years: the last ice age can be particularly clearly
seen in frame B, and the 'mini-ice age' of the late 17th century in Europe, in frame D.
    What are the causes of such changes, and how can we predict future variations?
Also, how can we predict the added effects of man's activities which are now

assuming global significance? These questions are currently being addressed by some of the best minds in science, and here we can do no more than give some overview of present ideas.

While there are many possible specific causes of the climatic fluctuations of the type seen in Fig. 2.7, we can define two quite distinct classes of causes:

(a) *External forcing*   External processes which can cause changes to the state of our climate, but which are unaffected themselves by the climatic state. Included in this category are changes in the orbit of the Earth, changes of solar output, continental drift, and even volcanic eruptions. These can all be thought of as external to the climate system.

(b) *Internal forcing*   Internal interactions and instabilities and feedback processes, which, even in the absence of any changes in the external forcing, can cause changes in the climatic state. Examples here are the incompletely understood phenomenon of the 'El Nino' (see Chapter 3), changes in the production or loss of greenhouse gases, and all of man's activities.

Many scientists have contributed to ideas in this field, amongst them being Lorenz, Budyko, and Saltzman. These people have worked particularly on the problem of understanding such variability, and thence arriving at the possibility of making some predictions. Lorenz (1963a, 1963b, 1968) examined the degree of determinism (or predictability) of the climate by considering the idea that there might be several quasi-stable states for the climate system, and that switches between these states might occur as a result of the natural internal variability and fluctuations of the climate system itself. He pointed out that what made this possible was the high degree of nonlinear interactions existing within the climate system. (We have already seen how the existence of nonlinear feedback processes can make the prediction of climate change so much more difficult than if processes are all linked in a simple, linear way). Lorenz went on to define the concept of an 'intransitive' system. This is a system for which, given randomly chosen initial conditions, there are two or more sets of long-term statistics, each of which has a greater than zero probability of occurring. In more simple terms, such a system could evolve along either of two or more paths. To even better describe the real Earth's climate, Lorenz then introduced the idea of an 'almost intransitive' system, in which there may be a single given solution extending over an infinite time interval, which will, nevertheless, possess successive, very long periods with markedly different sets of statistics. In other words, there may be only one path along which the climate can develop, but the state of the climate may be subject to sudden switches between metastable states within this one path.

Recent work by Hansen and co-workers (1988) may give some support to the idea of internally triggered variability, that is, changes between different states of the climate that might be triggered by internal 'noise' in the climate system. We shall see later that the explanation of the 'El Nino' phenomenon shows features that support this hypothesis. In calculations using a large three-dimensional coupled atmosphere–ocean model of the climate, Hansen and his group carried out a 100-year 'control' run, in which no changes of external forcing, nor of atmospheric composition, were introduced (the work was intended to examine the effects of increasing CO_2 and other greenhouse gases on the climate). These scientists have developed an exceedingly complex model, with many interactive and non-linear feedback processes built in, and what was observed in this control run was a sudden unforced — and unpredicted — change in global temperature of about 0.4 K. This occurred in one 20-year period near the middle of the a 100-year span. Though it would have to be shown that this was due to the complexity of processes simulated in the model, and not to some numerical vagary of the computer code or machine used, nevertheless, Hansen and his colleagues cite this result as supporting the view that unforced and unpredictable climate variability, triggered by internal fluctuations, may account for a large portion of historical climate change.

Saltzman (1983) remarks: 'We are dealing with an extremely complex, forced, dissipative system that undoubtedly contains a rich assortment of linear and nonlinear, positive and negative feedbacks'. He describes attempts to formulate a mathematical basis for understanding climate variability, which, though we will spare ourselves the pleasure of detailed examination, is worth describing here in outline. The climate is addressed by Saltzmann through the techniques of systems analysis, in which the time development of a climate parameter, say temperature T, is related to a set of other climate variables x_i, by

$$\frac{\partial T}{\partial t} = F(T, x_i, t) + R_1 + R_2 + \ldots + R_N \ , \tag{2.32}$$

where the function F is a formula usually obtained by a combination of theory and empirical analysis of data. The Rs are a set of 'forcing functions', i.e. a mathematical way of describing processes which are forcing a change on some climate variable such as temperature. The forcing functions may be either external (e.g. volcanic eruptions) or internal (e.g. melting of ice sheets). The name of the game is to limit the number of terms and interactions in equations like (2.32) to simplify the calculations, while retaining a sufficiently accurate description of reality.

Attempts are continuing to construct ever more realistic models, with recent efforts, for example, particularly going into a proper coupling of the atmosphere and the oceans. However, we are still some considerable way from a satisfactory capability to predict climate, if only to the limit defined by any inherent unpredictability of the climate system itself. In the meantime, it will become even more important to make accurate and long-term measurements of many climatic parameters on a global basis, and to carefully monitor any trends or changes. Clearly, satellite observations have a major role to play in this.

2.2 THE ATMOSPHERE

2.2.1 Composition and basic properties

The atmosphere of the Earth is a surprisingly thin and tenuous layer covering the surface of the planet, when seen from space. The atmospheric density decreases with height approximately exponentially, but processes of direct importance to the climate take place as high as 40 km, where the air pressure is as low as a few millibars (pressure at surface = 1013 mb). If converted to an effective thickness at surface temperature and pressure conditions, the atmosphere would be about 8 km thick. Compared to an Earth radius of 6370 km, the atmosphere therefore represents a relative thickness of about 0.1 per cent of this radius. Nevertheless, within this thin, fragile layer, virtually all of mankind's endeavours, sufferings, delights and ambitions are played out.

Table 2.3 gives some useful basic data on the atmosphere. The way that the

Table 2.3 — Physical and chemical properties of the atmosphere

Atomic and molecular parameters (dry air)

Boltzmann's constant, k	1.381×10^{-23} J K^{-1}
Planck's constant, h	6.6262×10^{-34} J s
Gas constant, R	8.3143 J K^{-1} mol^{-1}
Velocity of light, c	2.998×10^{8} m s^{-1}
Avogadro's number	6.022×10^{23} mol^{-1}
Loschmidt's number	2.687×10^{25} m^{-3}
Apparent molecular weight	28.964
Gas constant for dry air	287.05 J kg^{-1} K^{-1}
Specific heats for dry air at:	
constant pressure, cp	1005 J kg^{-1} K^{-1}
constant volume, cv	718 J kg^{-1} K^{-1}
Thermal conductivity	2.40×10^{-2} W m^{-1} K^{-1} at STP
Density of dry air at 273 K and 1013 mb	
pressure	1.293 kg m^{-3}

Properties of water vapour

Molecular weight	18.015
Latent heat of fusion at 273 K	3.34×10^{5} J kg^{-1}
Latent heat of vaporization at 273 K	2.50×10^{6} J kg^{-1}
Specific heat of liquid water at 273 K	4.218×10^{3} J kg^{-1} K^{-1}
Specific heat of ice at 273 K	2.106×10^{3} J kg^{-1} K^{-1}
Density of ice at 273 K	917 kg m^{-3}

Geophysical parameters

Earth's mean radius	6371 km
Standard surface gravity	9.80665 m s^{-2}
Standard surface pressure, p_0	1013.25 mb ($= 1.01325 \times 10^{5}$ Pa)
Ice point	273.15 K
Mean solar angular diameter	$31.99'$

density, ρ, or pressure, P, vary with height can be described by the hydrostatic equation,

$$dP = -\rho g \, dz \, , \tag{2.33}$$

where height is given by z and g is the acceleration due to gravity. For a perfect gas, the equation of state gives

$$\rho = \frac{MP}{RT},$$

(2.34)

where R is the molar gas constant.

Therefore, from equations (2.33) and (2.34) we obtain

$$\frac{dP}{P} = -\frac{dz}{H},$$

(2.35)

and integration gives

$$P = P_0 \exp\left(-\int_0^z \left(\frac{1}{H}\right) dz\right).$$

(2.36)

We can see that H represents a scale height, the height change over which the pressure changes by a factor of e. The value of H may be found from the above equations to be

$$H = \frac{RT}{Mg},$$

(2.37)

which is roughly 8.5 km under normal surface conditions.

The composition of the atmosphere is made up of a number of fairly constant major constituents, and a large number of minor constituents which make the chemistry of the atmosphere interesting, to say the least, and which give rise to the ozone layer, smog and other effects.

The composition of clean air at sea level is given in Table 2.4. The water vapour

Table 2.4 — Composition of the atmosphere

Constituent	Mixing ratio by volume
Nitrogen, N_2	0.78083
Oxygen, O_2	0.20947
Argon, Ar	0.00934
Carbon dioxide, CO_2	330×10^{-6} (increasing)
Neon, Ne	18.2×10^{-6}
Helium, He	5.2×10^{-6}
Krypton, Kr	1.1×10^{-6}
Xenon, Xe	0.1×10^{-6}
Hydrogen, H_2	0.5×10^{-6}
Methane, CH_4	2×10^{-6}
Nitrous oxide, N_2O	0.3×10^{-6}
Carbon monoxide, CO	0.1×10^{-6}
Water vapour, H_2O	Highly variable
Ozone, O_3	Highly variable

content is highly variable, ranging at the surface from about 10^{-2} in mixing ratio near the equator, to perhaps 10^{-3} at the surface near the poles, and falling to about a few times 10^{-6} in the stratosphere.

A number of 'model' atmospheres exist, extensive tabulations of pressure, density, temperature and to some extent composition, for a variety of locations around the Earth. These are useful 'standards' for the researcher to use in theoretical studies of the atmosphere. Since these tabulations are readily available in a number of publications, they will not be reproduced here, but can be accessed by the reader through any good library (e.g. the US Standard Atmosphere, 1976).

2.2.2 Spectral properties

We have already seen in earlier sections (2.1.6 and 2.1.7) how the molecular composition of the atmosphere leads to a complex absorption and emission spectrum at all wavelengths from the ultraviolet to the radio regions. This spectrum affects the radiation balance of the planet, and gives rise to a complex greenhouse effect, which we will study in more detail in Chapter 3.

The discussion given earlier, however, did not attempt to probe into the fine structure of the atmospheric spectrum, which we will now do. If the spectrum shown in Fig. 2.3 is expanded it reveals that, where a broad absorption band appears to occur, there is in reality a huge number of individual spectral lines due to a number of spectrally 'active' molecules. Table 2.5 lists the primary bands in the infrared portion

Table 2.5 — Primary infrared atmospheric absorption bands

Gas	Band centre (microns)	Doppler half-width at 300 K $(10^{-3} \mathrm{cm}^{-1})$	Lorentz half-width at STP (cm^{-1})
H_2O	2.7	6.4	0.11
	6.3	2.8	0.11
	20.0	0.9	0.11
	40.0	0.4	0.11
CO_2	4.3	2.6	0.15
	15.0	0.8	0.15
N_2O	7.8	1.5	0.16
O_3	4.7	2.3	0.16
	9.6	1.1	0.16
	14.1	0.8	0.16
CH_4	3.3	5.6	0.18
	7.7	2.4	0.18

of the spectrum: each one of these bands is composed of up to several thousand individual lines. A recent experiment on the Space Shuttle (the ATMOS project, to be described in more detail in Chapter 5) made measurements of the high-resolution spectrum of the stratosphere, looking at the Sun through the limb of the atmosphere (i.e. just above the horizon), and the results of that experiment beautifully demonstrate the complexity and richness of the spectrum. Some of the results obtained are illustrated in Fig. 2.8, which shows a region of the infrared spectrum between

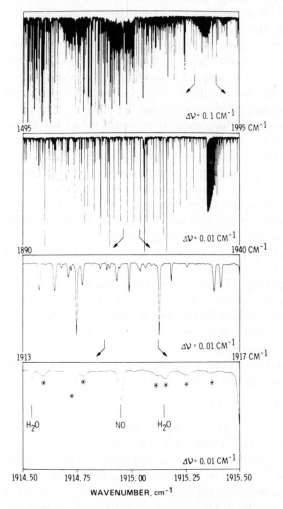

Fig. 2.8 — Part of the detailed infrared absorption spectrum of the atmosphere, in the region of 1900 cm^{-1} (5 microns wavelength), taken from the ATMOS experiment on Spacelab 2.

1495 cm^{-1} and 1995 cm^{-1}, and shows successive degrees of expansion of this spectrum to higher and higher resolution, eventually to the extent that the observed profiles of single spectral lines are revealed. We will show further examples of the detailed atmospheric spectrum in later chapters.

2.2.3 Chemistry
The chemistry of the atmosphere is also highly complicated. A great deal of work has been done over the last 10–20 years on the detailed chemistry of the stratosphere,

that part of the atmosphere lying between about 10 and 50 km, as a result of the recognition that pollution effects might be significant in this chemically very reactive region. Since the chemistry of the troposphere is not as well advanced (perhaps because it is, if anything, even more complex), and since it is particularly the stratosphere which is susceptible to study from space, we limit ourselves here to stratospheric chemistry only.

The stratosphere contains the ozone layer. As we shall describe in more detail later, in Chapter 3, ozone is a gas, a chemical derivative of the oxygen molecules in the air which we breathe. Instead of containing two oxygen atoms, as does molecular oxygen, O_2, ozone contains three atoms, O_3, and in consequence is highly reactive. Ozone is present in the troposphere around us, though in very small quantities, and is important in processes like smog formation. One of the most important roles of ozone is in the stratosphere, however. There, a dilute (few parts per million, or ppm) layer extends from about 15 to 40 km in altitude. This is really very rarefied (if compressed to the pressure and temperature at the Earth's surface, all the ozone in the stratosphere would be contained in a layer just 0.3 cm thick). Ozone in the stratosphere is important because it acts as a UV filter, cutting out shortwave UV radiation from the Sun which otherwise would be damaging to plants and animals as they have evolved on Earth. Also, as a result of absorbing this shortwave, highly energetic radiation, ozone causes a warming of the upper stratosphere, which influences global air motions.

The principal photochemical processes which form and destroy ozone in the stratosphere, and which thereby set up a dynamic equilibrium between formation and destruction, were first defined by Chapman over 50 years ago (Chapman, 1930). He showed that a series of reactions occurred, starting with the photodissociation of O_2 by sunlight at wavelengths below about 246 nm to form free oxygen atoms, which were then available to react with other oxygen molecules to form ozone. The key Chapman reaction in this formation of ozone is:

$$O + O_2 + M = O_3 + M \, , \tag{2.38}$$

where the symbol 'M' represents an 'air' molecule, i.e. predominantly a molecule of nitrogen or oxygen (see Table 2.4 for the composition of the atmosphere).

Other reactions which lead to the destruction of the ozone can occur, and taken together all the reactions which influence the density of ozone establish a dynamic equlibrium. The details of this equilibrium will become clearer in Chapter 3, when we look more closely into the photochemistry, but for now the important facts to recognize are that this equilibrium exists, and that any changes to the concentrations of the various chemical species that play a part in the balance can cause a shift in that balance. At the present time, this dynamic equlibrium leads to a layered distribution of ozone in the stratosphere (see Fig. 3.1 in Chapter 3).

In the 1970s, through pioneering work by Paul Crutzen, Harold Johnston and others (e.g. see Crutzen, 1970), it was realized that in order to reproduce the observed concentrations of ozone quantitatively, other chemical processes than those involving purely oxygen photochemistry must be taking place, because the amounts of ozone predicted by this theory were considerably too high. These scientists introduced the idea that ozone destruction could also be taking place through reactions involving other trace chemicals, present in much lower concentrations than the ozone itself. The basis of this suggestion rested on the idea of 'catalysis', which many of us learn about in school chemistry. In catalysis, a trace constituent can destroy or convert a chemical present in much higher concentrations if it is used and subsequently re-created chemically: thus, if this cycle is repeated many times, the trace constituent can in the limit eventually 'eat up' all of the more dominant chemical. The important breakthrough came when it was realized that other chemicals in the atmosphere, such as the nitrogen oxides, NO and NO_2 (albeit present in concentrations 1000 times lower than ozone itself), could create catalytic cycles which lead to ozone destruction on a significant scale. NO and NO_2, present in the stratosphere in concentrations in the 10^{-9} to 10^{-8} range could affect the balance of ozone concentrations (which occur in the 10^{-6} to 10^{-5} range) through the following combination of reactions:

$$NO + O_3 = NO_2 + O_2 \qquad \text{(a)}$$
$$NO_2 + O = NO + O_2 \qquad \text{(b)} \qquad\qquad (2.39)$$
$$\text{Net:} \qquad O + O_3 = 2O_2 \qquad\qquad \text{(c)}$$

The net effect of this catalytic cycle is that O_3 is reduced to O_2, and the NO and NO_2 are constantly recycled, but not destroyed, i.e. they act as catalysts. Later work has revealed that chlorine Cl, and chlorine monoxide, ClO, can play a similar role in ozone destruction, as can OH and HO_2. Modern theories of ozone in the stratosphere have become very complex, with over 100 identifiable chemical reactions having to be progressed simultaneously as the computer models that have been developed step forward in time (see Brasseur and Solomon, 1984).

In Chapter 3 we will take this story of ozone in the stratosphere a stage further, when we come to consider the question of the Antarctic ozone hole. In Chapters 4 and 5 we will consider how modern space instrumentation can make global measurements which are critical to our understanding of stratospheric chemistry.

2.2.4 Dynamics

The dynamics of the atmosphere also presents us with numerous complexities. We are all well aware of the variability of winds, storms, fronts and so on. Basically winds are caused by differences in pressure at different places, and these differences in pressure originate in the asymmetric heating caused by latitudinal and diurnal variations of solar input, seasonal effects, surface topography and roughness, and so on. This heating creates areas of high and low pressure, and air moves from one to the other to equalize the pressure. The actual motion is not directly between the two centres of high and low pressure, however, because of the effects of the spinning of the Earth, which induces an apparent force called the Coriolis force; this causes a moving parcel of air to veer to the right of the expected track in the northern hemisphere (NH), and to the left in the southern hemisphere (SH), simply because the Earth is spinning below the moving air. This gives rise to an anticlockwise circulation around low pressure areas in the NH, and clockwise around centres of high pressure; again, in the SH the opposites apply.

It is interesting to explore why this is so, in simple mathematical terms. Since the atmosphere is so thin compared with its horizontal extent, we may apply a quasi-horizontal approximation. Then the vector forces acting on an air parcel, ignoring the effect of gravity (which does not affect the quasi-horizontal problem), are:

— friction with the surface $\mathbf{F}$

— pressure gradient force $\dfrac{1}{\rho}\nabla P$ (2.40)

— Coriolis force $2\Omega\sin\phi\ \mathbf{v}\wedge\mathbf{k}$.

Newton's second law of motion gives

$$\frac{d\mathbf{v}}{dt}=f\mathbf{v}\wedge\mathbf{k}-\frac{1}{\rho}\nabla P+\mathbf{F}$$ (2.41)

where $f=2\Omega\sin\phi$
         $\phi=$ latitude
         $\mathbf{k}=$ unit vector in vertical direction (z).

If we consider the so-called geostrophic approximation, which is widely used and which simply means that we ignore the effects of friction, everywhere except at the lowest surface layers of the atmosphere, or,

$$\mathbf{F}=0\ ,$$

and we assume constant motion (no acceleration), then

$$\frac{d\mathbf{v}}{dt}=0\ ,$$

and we find that the pressure gradient force is balanced by the Coriolis force,

$$f\mathbf{v}\wedge\mathbf{k}=\frac{1}{\rho}\nabla P\ .$$ (2.42)

From the properties of vector multiplication, $\mathbf{v}$ is at right angles to the $\mathbf{k}$ vector, and also at right angles to the direction of ∇P, and in fact is to the right of the direction of ∇P in the NH and to the left in the SH. It is interesting to note that the geostrophic approximation is valid only well away from equatorial regions (where the Coriolis force in the horizontal is virtually zero) and above the boundary layer (say $z=1$ km), where $\mathbf{F}$ can be ignored.

   Fig. 2.9 illustrates the geostrophic flow around a cyclonic low pressure area in the NH. We can also see why a tighter clustering of isobars on the weather man's chart means stronger winds — since $\mathbf{v}\wedge\mathbf{k}$ must balance ∇P, and if ∇P is greater (as indicated by tight bunching of the isobars), then so $\mathbf{v}$ must be greater.

2.3 CLOUDS

2.3.1 Clouds and radiation

One other major component of the atmosphere system is cloud. As can be witnessed from any photograph of the Earth from space (see Plate 1), the globe is swathed in a

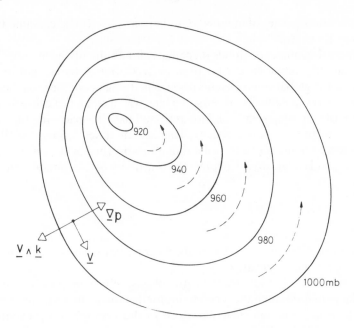

Fig. 2.9 — Illustrating geostrophic flow: the terms are defined in the text.

constantly changing film of clouds, looking very thin but very beautiful. It would seem perhaps obvious from their all-pervading presence in the atmosphere that clouds must play an important role in the climate system, but as we shall see, our understanding of the details of that role is surprisingly poor. Indeed, scientists have been unclear whether the effect of clouds on the climate system is to produce a net cooling or a net heating: for some time, evidence seemed to be pointing to there being an almost perfect balance between these competing effects.

Clouds are formed by the condensation or freezing of water vapour onto condensation nuclei such as dust particles. It is not our purpose here to go into the physics of cloud formation, interesting though that is (for example, see Mason, 1974), nor into their various types, though it is clear that we can form a basic classification scheme depending on whether the clouds are high, medium or low, formed of ice or water, and on their horizontal and vertical structure.

For our purposes, clouds interact with the radiation field in the atmosphere, and so modulate both the incoming stream of solar radiation and the outgoing stream of thermal radiation. They must clearly have a strong influence on the local radiation and thermal balance of the atmosphere and the underlying surface, but what of their globally averaged effects? Also they have an important relevance to the study of the climate system from space. Clouds are opaque in the visible and the infrared: however, at wavelengths longer than a few millimetres, clouds become more and more transparent to electromagnetic radiation. Thus to measure the radiative temperature of cloud tops, for example, we should best employ infrared techniques, whereas to measure, say, ocean surface temperature below persistent cloud, we should utilize the properties of microwave radiation. This will be examined in more detail later on.

To return to the impact of clouds on climate, it has already been noted that the state of our understanding is very far from perfect, though the advent of satellite observations of the global cloud systems has clarified a lot of our ideas on the subject. The effect of cloud can be two-fold. First, an increase in cloud can cause an increase in the amount of incoming solar radiation reflected back to space, in other words, a loss of energy to the terrestrial system. This is sometimes termed the albedo or shortwave effect of cloud: if Q is the solar radiation absorbed by the system, and A_c is the fractional cloud cover, we can express the sensitivity of the climate system to this effect by the ratio

$$\frac{\partial Q}{\partial A_c}, \tag{2.43}$$

where, by definition, because Q refers to the solar radiation absorbed by the system, $\partial Q/\partial A_c$ must be negative in value.

The second effect, the longwave or greenhouse effect of clouds, can cause a heating of the global system if the increase in cloud amount causes an increase in the amount of thermal radiation that is trapped in the system, and prevented from escaping to space. If F is the outgoing longwave flux of radiation at the top of the atmosphere, this effect can be expressed in a similar ratio to that in equation (2.43), as in

$$-\frac{\partial F}{\partial A_c}, \tag{2.44}$$

where in this case the negative sign indicates that an increase in cloud amount has an opposite effect on F from the effect of the same increase on the solar radiation absorbed by the system, Q. We should note, however, that again because of the definition of F as the longwave flux at the top of the atmosphere, the term $\partial F/\partial A_c$ must also be negative in value. We can define a climate sensitivity parameter, δ, by

$$\delta = \frac{\partial(\text{Net})}{\partial A_c} = \frac{\partial Q}{\partial A_c} - \frac{\partial F}{\partial A_c}, \tag{2.45}$$

where we have written 'Net' for the difference

$$\text{Net} = Q - F. \tag{2.46}$$

Clearly, if the first term in equation (2.45) dominates, (i.e. δ is negative in value) then we can say that the shortwave effect is stronger than the longwave effect, and the effect of an increase in cloud amount is to cause a net cooling of the planet. If, however, the longwave term in equation (2.45) dominates, (i.e. the value of δ is positive) then the opposite applies, and an increase in cloud amount causes a net heating of the planet because less longwave radiation is allowed to escape to space. We might ask which is the true situation? The fascinating thing we find is that it has proven surprisingly hard to establish whether the climatic effect of clouds globally is to cause net heating or net cooling, or even if the effect is exactly zero, that is the longwave and shortwave effects perfectly cancel. We will now examine what both theoretical and experimental studies of this question have taught us, and we will see what an important role satellite data have played in this story. (The reader may find a valuable, more extensive review of the subject in Ohring and Gruber (1983), and in Arking (1989)).

2.3.2 Model results

Theoretical studies of the effect of clouds on the radiation field require the development of models which include quite complex treatments of the radiation transfer processes within the atmosphere, and assumptions about the cloud amount and distribution, both horizontally and with height, the temperature and water vapour distributions, and feedback and other interactive processes with the surface, space, etc. These effects cannot be exactly represented in a computer model, for reasons of the finite size of the calculations possible in even the largest computer, as well as limitations in our understanding of various processes involved; so the different models produced by different scientists give differing results. The work that has been done covers models which simulate the cloud effects in 1-, 2-, and 3-dimensions of space, and variously include interactions with the land surface, oceans, ice, etc. The resultant calculations of the term $\partial F/\partial A_c$, for example, cover a range of about a factor of 2, owing to the uncertainties and differences outlined above. However, up to about 1980, when new data sets became available from satellite programmes, the various theories agreed generally that values of δ were at least negative, i.e. that the cloud albedo effect dominates. This is shown by the results presented in Table 2.6, taken from a variety of sources.

Table 2.6 — Calculated cloud–climate sensitivity parameter

Method	Reference	$\partial Q/\partial A_c$ (W/m^2)	$\partial F/\partial A_c$ (W/m^2)	δ
1-D, global average model:				
(single cloud)	Ramanathan, 1976		− 71	− 24
(multiple clouds)	Cess, 1974		− 68	− 27
(multiple clouds zonal average)	Hoyt, 1976	− 95 ± 10	− 34	− 61
2-D, zonal average, single cloud	Ohring and Adler, 1978		− 33	− 62
3-D, global circulation model	Coakley and Wielicki, 1979		− 71	− 24

2.3.3 Satellite data

This rather tidy situation as obtained by the modellers was somewhat complicated in the early 1980s, by new observational evidence! The new facts came from the analysis of satellite-derived data sets from the Nimbus and NOAA satellite series (see Chapters 4 and 5). Cess (1976), for example, used satellite measurements of zonal, annually averaged values of outgoing longwave radiation, together with conventional sources of data on cloudiness and surface temperature, and obtained a value of $-91\,\mathrm{W\,m^{-2}}$ for $\partial F/\partial A_c$, and a value of $-2.6\,\mathrm{W\,m^{-2}}$ in the northern hemisphere and $0.6\,\mathrm{W\,m^{-2}}$ in the SH for δ, i.e. essentially zero within the uncertainties of the calculations. This implied that the balance between the albedo and the cloud greenhouse effects is virtually perfect, at least for the typical cloud conditions presently on the Earth (one should be careful to recognize that such a result might no longer apply to a system which is pushed a long way out of equilibrium). This was an intriguing result, of course, because it implied a fantastic degree of balance between the competing heating and cooling effects of clouds: this in turn would mean that the climate, at least with respect to perturbations of clouds, was in a state of equilibrium, which perhaps registered positively with those scientists inclined to a view that the Earth's climate was highly self-stabilizing.

However, the controversy did not stop there, because similar estimates by other authors, such as Ohring and co-workers, have yielded different results from those of Cess, in these cases showing that values of δ based on satellite data are negative over most of the globe, indicating dominance by the albedo effect, as was predicted by the modellers. Table 2.7 includes some of the results based on satellite data, illustrating

Table 2.7 — Measured cloud–climate sensitivity parameter

Method	Reference	e
Analysis of outgoing flux and albedo	Ohring et al., 1981	0.33
	Hartman and Short, 1980	0.4
Direct analysis of satellite data	Cess et al., 1982 (using various satellite data sets: see Ohring and Gruber, 1983)	1.1 1.0 0.5 0.51

$$e = \frac{\partial F/\partial A_c}{\partial Q/\partial A_c} = \partial F/\partial Q$$

($e < 1$, cloud albedo dominates; $e > 1$, cloud greenhouse dominates)

the spread of results. As possible reasons for these discrepancies, some authors have noted certain limitations in the observational data, which might lead to the differences reported (for example, the satellite data used in these particular studies were from narrow-band infrared radiometers on spacecraft, and did not measure the entire spectrum of radiation; and also the observations may not be able to account adequately for the effects of low clouds underlying higher clouds). Also, it is clear that the treatment of all the important processes within the available computer models is far from perfect.

Later space missions (see Chapter 4) have included measurements across a much broader band of radiation, and have therefore involved fewer assumptions about the extrapolation of the earlier narrow-band data to all wavelengths. For example, Ramanathan and co-workers (1989) used data from the Earth Radiation Budget Experiment (ERBE), which flew on the Nimbus 7 spacecraft, launched in 1978, to produce broad-band radiation fluxes. They deduced the values of parameters called C_{LW} and C_{SW}, representing the longwave and shortwave (essentially the infrared and visible) components of the thermal and solar fluxes at the top of the atmosphere. These are defined as the values of the parameters $\partial F/\partial A_c$ and $\partial Q/\partial A_c$, integrated over the cloud amount, A_c.

$$C_{LW} = -\int_0^{A_c} \left(\frac{\partial F}{\partial A_c}\right) dA_c = -\bar{A}_t\left(\frac{\partial F}{\partial A_c}\right) \tag{2.47}$$

$$= F(0) - F(\bar{A}_c)$$

$$C_{SW} = \int_0^{A_c} \left(\frac{\partial Q}{\partial A_c}\right) dA_c = \bar{A}_c\left(\frac{\partial Q}{\partial A_c}\right) \tag{2.48}$$

$$= Q(\bar{A}_c) - Q(0) .$$

In these equations, the term $\bar{A}_c$ represents the average over the complete range of time and space represented by the measurements. The two parameters C_{LW} and C_{SW} represent the total effect of cloud cover on the longwave and shortwave heating in the atmosphere, called 'cloud forcing' by Ramanathan. The net 'forcing' by clouds can be related back to the parameter δ by the equation

$$C = C_{LW} + C_{SW} = \bar{A}_c\delta . \tag{2.49}$$

Cess (1976) and Ramanathan *et al.* (1989) obtained the following values for these parameters (units W/m^2):

	Cess	Ramanathan
C_{LW}	45.5	31
C_{SW}	− 44.5	− 48
C	+ 1	− 17

Thus, substantial differences still remain, even between analyses based on observational data, as well as between theoretical treatments. Arking (1989) notes that all of the analyses, except those based on complete spectral observations of the actual total integrated fluxes at the top of the atmosphere (e.g., the ERBE data), assume that the relationship between flux and cloud cover is linear. It would seem that different types of cloud have different values of sensitivity parameter, and so in all but the simplest of cases, the derivatives $\partial Q/\partial A_c$ and $\partial F/\partial A_c$ will be functions of

A_c. Indeed, intuitively, it would seem improbable that a simple linear relationship would hold, considering the complexity and variations of most typical cloud scenes.

Therefore, it is only possible to conclude that the weight of evidence seems to indicate that the true global effect of clouds lies somewhere between a perfect balance between the albedo and greenhouse effects, and a moderately strong domination of the albedo effect. The study of the problem continues, and no doubt accuracy will increase with time in this very complicated but extremely important problem. In his recent survey of the problem for the International Radiation Commission, Arking (1989) concludes that

(1) clouds may have a very strong influence on climate change; and that
(2) we are far from knowing the magnitude of this influence.

2.4 THE OCEANS

2.4.1 Basic properties

The oceans are an important component of the climate system. The heat capacity of water is large, so that the quantity of heat stored in the oceans is enormous, and acts as a kind of thermal 'buffer' — or reservoir — for the whole climate system. Moreover, as we discovered earlier in this chapter, the large thermal capacity leads to a rather long characteristic time constant: in other words, the oceans take a long time to heat up, and also to cool, so that they tend to moderate temperature swings. As a consequence, as is well known in geography, coastal climates are perhaps damper but are more equable than mid-continental climates, which can exhibit huge summer-to-winter swings of temperature.

The primary parameters which describe the properties of sea water are:

> — density
> — temperature
> — salinity.   (2.50)

The relationship between the three parameters (only two of which are mathematically independent) is complex and is illustrated in Fig. 2.10. A typical temperature profile in the open ocean is shown in Fig. 2.11. Generally, with few exceptions, the temperature of the oceans decreases with increasing depth. At the top of the ocean there exists a mixed layer a few tens to a hundred metres thick, which is stirred by surface winds. Below this mixed layer the thermocline is a region of rapidly falling temperature, which shows some variability depending on the season. At depths below the thermocline, the oceans become slowly colder, until they become virtually isothermal, at temperatures below about 5°C. The low temperature of deep open ocean water indicates that it might originate at high latitudes via a global circulation.

The area, volume and mean depth of the major oceans (including adjacent seas) is given in Table 2.8. At a density of about $1\,\mathrm{g\,cm^{-3}}$ the world's oceans therefore

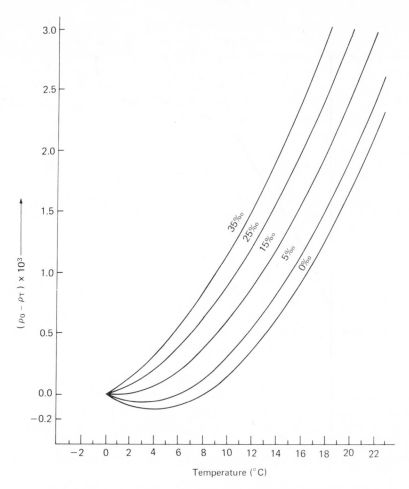

Fig. 2.10 — Density of pure water and of seawater of different salinities as a function of temperature: ρ_0 is the density at 0°C, and ρ_T the density at T°C (units g cm^{-3}).

weigh something like 10^{21} kg, or a billion billion metric tonnes. By comparison, the atmosphere weighs some 4×10^{18} kg, or 4 million billion metric tonnes.

Recently, attention has been given to the possibility of direct thermal expansion of the oceans, resulting from global warming, giving rise to increases in sea level. It is now felt that this might be a more immediate effect than, for example, the melting of the large ice sheets of Antarctica and Greenland, which would have much longer thermal time constants than would oceanic expansion. Estimates of the sea level rise which has occurred during the last 100 years fall in the range 10–20 cm (Gornitz *et al.*, 1982) and of this, it is calculated that between 2 and 6 cm could be due to direct thermal expansion (the other significant sources being the melting of glaciers, which respond quite fast to temperature changes, and of the Greenland ice sheet). Interestingly, current thinking is that the Antarctic ice sheet has contributed rather

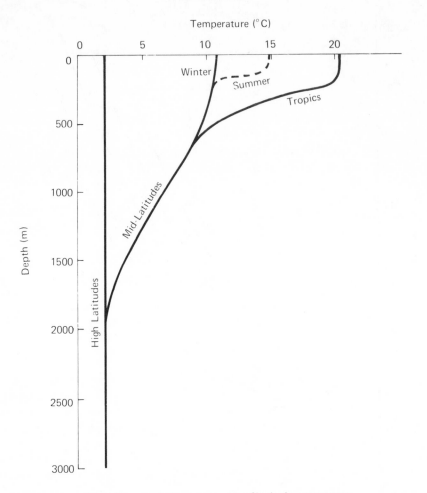

Fig. 2.11 — Typical temperature profiles in the open ocean.

Table 2.8 — Area, volume and mean depth of oceans

Ocean	Area (10⁶ km²)	Volume (10⁶ km³)	Mean depth (m)
Atlantic	106.463	354.679	3332
Pacific	179.679	723.699	4028
Indian	74.917	291.945	3897
Total	361.059	1370.323	3795 (on average)

little to the volume of water, because any melting seems to have been balanced by increased precipitation: however, there are very large uncertainties attached to many of the data on which these conclusions are based, particularly with regard to the Antarctic.

2.4.2   Heat storage and energy balance

The rate at which heat is gained or lost by the surface layer of the ocean is given by

$$q_M = q_s - q_e - q_v - q_c \, , \tag{2.51}$$

where q represents a flux of energy across a unit area of sea surface, and the subscripts refer to

M = mixed layer energy gain or loss,
s = solar insolation (heat gain),
e = thermal emission (heat loss),
v = evaporation (heat loss),
c = convection and conduction (heat loss).

The equation may be out of balance locally, but globally and over a sufficient period of time (e.g. ~ 1 year), it must be in balance. The net source of energy is by solar insolation, predominantly in equatorial regions, balanced by various distribution and loss mechanisms at higher latitudes.

The various terms are highly dependent on variables such as wind stress, cloud cover, sea ice amount, etc. Thus it is not possible to develop any simple analytical expressions to describe the balances. Fig. 2.12 shows a schematic diagram which gives approximate percentage figures for the atmosphere-ocean heat balance, averaged over latitude and season.

The diagram in Fig. 2.12 refers to the Earth as a whole, but does illustrate some interesting aspects of the relative influences of the oceans, land and atmosphere. First, we can see that the shortwave albedo of the total system is about 30 per cent (see previous section on clouds). Also, we note that the amount of solar energy directly absorbed by the atmosphere is low (about 19 per cent), compared with the 45 per cent transferred from the ocean and land surface to the atmosphere. Because of their much greater thermal capacity (see Table 2.1) the oceans are much more effective than the land as a thermal store, absorbing heat during the summer and releasing it during the winter, and thereby reducing extremes of climatic variation, as we saw above.

Returning to Fig. 2.12 we should note that, while the diagram is an all-Earth average, the oceans alone are dominated by evaporation (latent heat) exchange with the atmosphere, accounting for some 2/3 of the energy exchange, with longwave radiation and sensible heat exchange (conduction and convection) less important than over land.

2.4.3   Dynamics

Earlier in this chapter, we wrote down an equation which described the forces acting on a single air mass in a plane parallel stratified atmosphere. An analogous equation

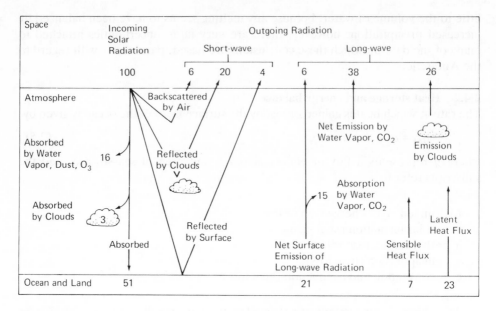

Fig. 2.12 — Ocean–atmosphere heat balance. 100 units of incoming solar irradiance are scattered, reflected and absorbed in the proportions shown. Values are whole-Earth yearly averages.

can be written down for the oceans, expressed as Newton's second law of motion, summed over all forces, F_i, acting in the horizontal

$$\rho\frac{d\mathbf{v}}{dt} = \sum_i F_i \, . \tag{2.52}$$

Expanding we get

$$\frac{d\mathbf{v}}{dt} = -\frac{1}{\rho}\boldsymbol{\nabla}P + f\mathbf{v}\wedge\mathbf{k} + \mathbf{F} \, . \tag{2.53}$$

Equation (2.53) excludes gravity, which does not of course affect horizontal motions, and expresses the sum of all the other forces acting on a given water mass. The frictional force $\mathbf{F}$ can be separated into two components, that due to the wind stress, W_{xy},

$$\mathbf{F}_w = \frac{1}{\rho}\boldsymbol{\nabla}W_{xy} \, . \tag{2.54}$$

plus the internal friction of the the upper ocean, which is dependent on velocity, $\mathbf{v}$,

$$\mathbf{F}_i = J\mathbf{v} \, , \tag{2.55}$$

where J is a constant which depends on parameters such as molecular viscosity.

Thus, broadly, the dynamics of the ocean resembles that of the atmosphere, except that, because of the greater density, ρ, motions are much slower, whereas (on the other hand), because of the greater thermal capacity, greater quantities of heat are moved around for a given velocity than in the atmosphere. In fact, the motion of the ocean in re-distributing heat from equatorial regions to the poles transports roughly the same amount of energy as is transported by the atmosphere from the equator to the poles.

The importance of the dynamics of the oceans to the global energy balance and the climate is illustrated by Figs 2.13 and 2.14. Fig. 2.13 shows the major current systems of the world's oceans, and Fig. 2.14 shows the heat flux from the ocean surface to the underlying layers of water, due to horizontal heat exchange. This clearly illustrates the role of advection of energy by the oceanic currents shown in Fig. 2.13, in redistributing heat from equatorial to mid-latitudes and polar regions.

2.4.4 Spectral properties of the oceans

The principal concern which we have from the perspective of remote sensing of ocean/climate parameters from space is over the values of the reflectivity and emissivity of the ocean surface at all the wavelengths which might be employed from space. As is well known from basic optical theory (see equation (2.28)), the reflectivity R of a dielectric surface under normal incidence, and assuming that the refractive index of air is unity, is given by

$$R = \frac{(n-1)^2 + \kappa^2}{(n+1)^2 + \kappa^2}, \tag{2.55}$$

where n is the refractive index of the water. Knowing that reflectivity from an optically thick surface is equal to one minus emissivity (see equation (2.13)), and knowing (from standard tables) the values of n for sea water at different wavelengths, we can estimate R and ε as follows

Waveband	R	ε
Visible–infrared	0.02–0.01	0.98–0.99
Microwave (10 GHz)	~0.61	~0.39

Fig. 2.15 shows a plot of the reflectivity (or 'reflectance') against angle of incidence of the radiation on the ocean surface for two values of radiation wavelength, one in the visible region at 0.55 microns, and the other in the microwave range at 3 cm. So far we have discussed only the normal incidence case, which corresponds to the points on the y-axis in Fig. 2.15. However, we see that in general the reflectivity increases with increasing angle of incidence, reaching an asymptotic value of 1.0 at 90 degrees, of course. The figure shows data for both horizontal and vertical polarization of the incident radiation (polarization refers to an intrinsic

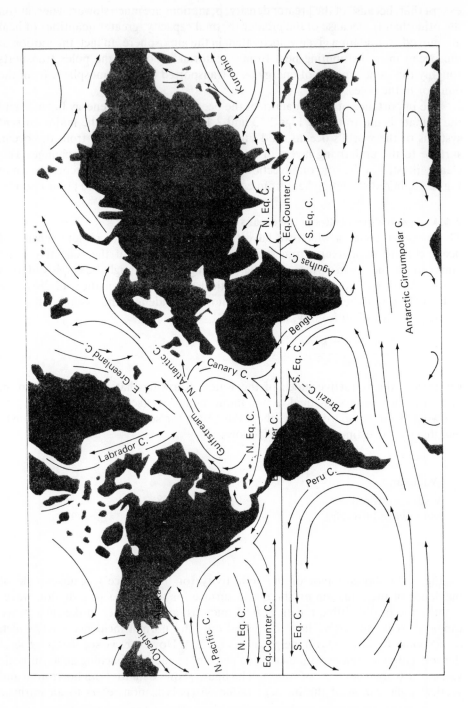

Fig. 2.13 — Major surface ocean currents of the world.

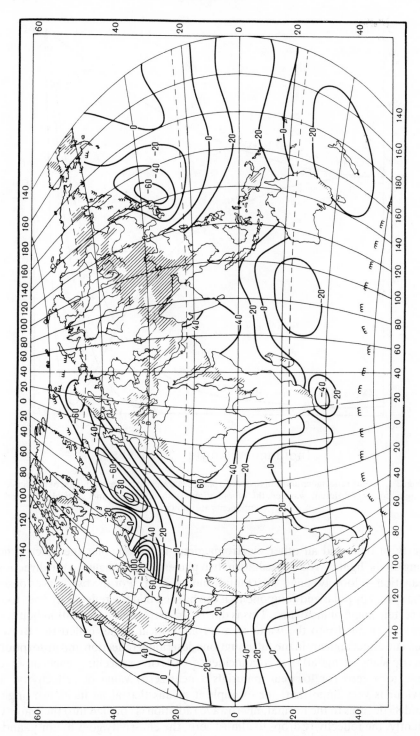

Fig. 2.14 — Heat flux from the ocean surface to the underlying layers of water (units: $kcal\,cm^{-2}\,yr^{-1}$).

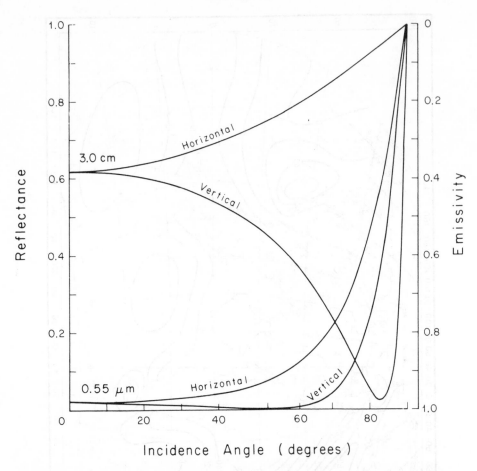

Fig. 2.15—The reflectance and emissivity of a plane sea water surface as a function of incidence angle, for two different wavelengths of radiation. Both horizontal and vertical polarized components are shown.

property of light and all other electromagnetic radiation, that the electric fields making up the light wave can oscillate in two planes at right angles to the direction of propagation — loosely called 'horizontal' and 'vertical'. For the two curves describing the vertically polarized component, it can be seen that the reflectivity goes to zero for a certain angle of incidence. This is due to this particular polarization being very effectively coupled into the water at a certain critical angle, called the Brewster angle. The effect appears in the full equations describing reflection and absorption by dielectric slabs of material as a coincidence of terms cancelling each other out, leading to a zero (or for real materials a near-zero) value of reflectivity. This behaviour is very important: for example it means that at an incidence angle of around 85 degrees to the normal, microwave radiation is coupled into the water very efficiently, the reflectivity drops to almost zero, the emissivity goes to unity, and the incident radiation is effectively all absorbed into the ocean. A similar effect occurs in the visible, but since the result is only to decrease the reflectivity from about 0.02 to 0.0, the differential influence is small. In the microwave, however, the reflectivity falls from about 0.6 at normal incidence to about 0.02 at the Brewster angle.

To conclude this section on the spectral properties of sea water, we show Fig. 2.16, as a useful reference. It shows how the reflectivity varies with wavelength

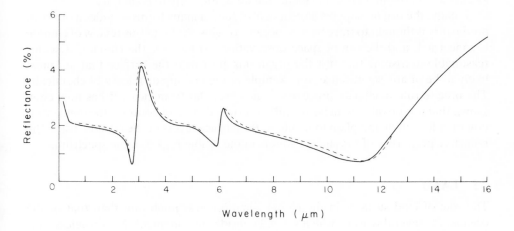

Fig. 2.16 — The reflectance spectrum in the infrared of pure water (solid line) and seawater (broken line).

throughout the visible and the infrared, before beginning its rise at long wavelengths to values of about 60 per cent that we have seen occur at microwavelengths. The sharp edges observed at about 3 and 6 microns correspond with intense absorption bands of water at these wavelengths.

2.4.5 Chemical effects, aerosols, and the biosphere

The oceans have other important effects on the climate. Thus the vast open water surface, covering 2/3 of the Earth's surface area, represents a huge 'solvent' for dissolving and absorbing gases in the atmosphere. There is a great deal of uncertainty surrounding the quantitative aspects of such processes — for example, how much carbon dioxide (CO_2) can the oceans absorb before they become saturated, and unable to absorb any more? What is the effect of/on lifeforms in the ocean? etc. Thus a complex combination of physical, chemical, biological and geological processes is involved, and our understanding is far from complete. It is known, however, that atmospheric concentrations of carbon dioxide have increased in direct proportion to global production by fossil fuel burning over the past century, and it appears, from a comparison of the production on the one hand and the increases observed in the atmosphere on the other, that approximately 1/2 of the production is unaccounted for, and presumably has been taken up by the oceans. Very recent work seems to show that a significant part of the oceanic take-up of CO_2 is due to absorption by microorganisms in the oceans.

The oceans can also affect the atmosphere, through the production of marine aerosol, that is droplets of water and brine solution thrown into the air by wave-breaking, or evaporation and condensation. Such suspended particulates can cause local heating of the air by absorption of solar radiation, and slight cooling of the underlying ocean. However, it is likely that the effects of marine aerosols are fairly localized, and the influence on global climate is not likely to be major.

Finally, the oceans support a great deal of life, in many forms — indeed our own pre-history is thought to trace back to oceanic origins. As far as this review of climate is concerned, and the use of space observations to monitor the climate, it seems reasonable to assume that it is the organisms at or near the surface that are most likely to be of any significance, for example in the take-up or release of chemicals. The precise role in climate processes is, however, far from clear. It has now been shown that it is possible to measure with a certain degree of accuracy the concentration of chlorophyll and planktonic life in the surface layers of the oceans, using the reflective properties of the surface waters in the visible region of the spectrum.

2.5   LAND

The role of land surfaces in the global climate is less important than that of the oceans, for several reasons which we will consider in a moment. Nevertheless, it is not possible to ignore land-surface processes, and it is also important to remember that the impact of climatic processes on the land may be a very direct manifestation of climate change, and one which can have very direct and dire consequences for the people who suffer the effects of drought, flood, or desertification. The reasons why the climatic importance of the land is not so great as that of the oceans are several. The thermal capacity of land is less than that of water (see Table 2.1), and so the amount of thermal 'storage' is less; there is no possibility of overturning or circulatory motions to redistribute energy globally; and since land is not always moist, the role of latent heat processes — so important in ocean–atmosphere exchange processes — is less extreme than for the oceans. Also, the thermal/radiative effects do not penetrate as deeply as in the partially transparent oceans.

Despite these points, as we have already seen, the processes at work in the climate system which involve the land must be taken into account. As well as the disastrous phenomena mentioned above, issues such as rainfall, crop production, aridity, river and lake supply and run-off, are all important on both regional and global scales.

2.5.1   Heat capacity and temperature variations
We know that because of its lower heat capacity, lack of dynamical motions, lack of transparency in depth, and also because of its lower capacity for latent heat release, land is a poorer store of heat energy than the oceans. For these same reasons, a given input or loss of heat energy to a volume of land (soil, sand, etc.) will give rise to a greater increase or fall in temperature than is the case for the ocean. (This is a consequence of the simple relationship between mass, temperature and thermal mass described by equation (2.1) earlier.) Therefore, in the global climate, the land

surface acts less as a stabilizing and moderating influence: in response to seasonal and even diurnal changes of solar insolation, the land surface can undergo a dramatic change in temperature. For example, under the dry, anticyclonic conditions prevailing over most desert areas, extremely high daytime temperatures of over 40°C can be followed during the clear night by a ground frost, in other words a fall in temperature of about 40°C. Indeed, the dew that forms during the night in such circumstances can be the sole source of moisture for certain forms of plant life, well adapted to the prevailing harsh and extreme conditions.

Fig. 2.17 illustrates this poor thermal capacity of the land surface: it shows the

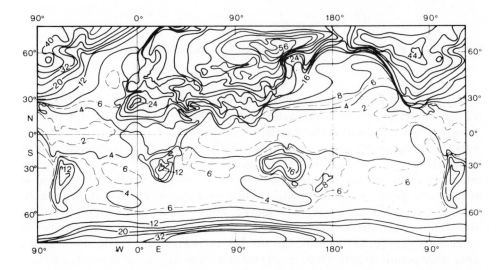

Fig. 2.17 — The annual range of land-surface temperatures.

annual range of surface temperatures globally. From this it can be seen that oceanic temperature ranges of 0°C to perhaps 10°C compare with mid-continental and desert ranges of perhaps up to 50°C and more. In later chapters we will show some satellite measurements of the outgoing infrared radiance from a variety of surfaces and conditions, which will illustrate these differences.

2.5.2 Models of land surface processes

Theoreticians have developed models of the land-surface processes important to climate, and present efforts are being directed towards coupling these models with global models of the atmosphere and oceans. The radiation balance at the land surface includes the same inventory of terms which apply also to the ocean and the atmosphere, though, of course, with different degrees of importance. The two extreme cases to model in simple terms are the dry-soil and wet-soil cases. Fig. 2.18 illustrates such models, using an example taken from the work of Mintz (1984), and defines the separate terms in the balance. The length of the arrows in the diagram indicates the size of each term: the most obvious difference being in the latent heat

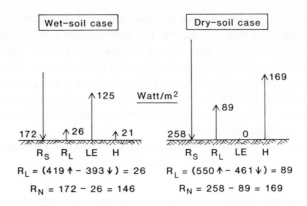

Fig. 2.18 — Dry-soil and wet-soil models of energy exchange with the atmosphere.

energy release (LE) owing to evaporative cooling of the ground, and in direct heat transfer (H). The enormous difference made to the global surface energy balance owing to the dampness of the land surface is shown in Fig. 2.19, taken from the same work. In this figure, which shows the ground surface temperature calculated for both wet-soil (top) and dry-soil (bottom) conditions, we can see that the wet-soil case is, rather obviously, tending towards a 'global ocean' model. The calculations used an atmospheric global circulation model, so that some degree of sophistication in the associated feedback processes was achieved. For example, in the dry-soil case the model indicates much less cloudiness, which allows a larger amount of solar radiation to reach the ground, as is indicated by the R_s term in Fig. 2.18. The resultant higher surface temperature also leads to greater longwave emission in the dry-soil case (R_L). The model calculations for these two extreme cases also yielded interesting and instructive differences in atmospheric circulation patterns, which we do not have space to discuss in detail here, but which obviously take us further along the path of understanding in detail how and why variation in local climates occur, variations which can have such devastating effects on mankind (drought in marginal desertic zones in Africa, floods in monsoon zones such as Bangladesh, drought in northern continental zones such as the USA and China, and so on). Such results also help us to appreciate that although the land is not such a dominant term in the global climate 'equation' as is the ocean, it is nevertheless important.

Clearly, satellite observations have a great deal to offer in developing our comprehension of the interaction of the land surface with the climate system: we shall return to this theme in later chapters.

2.5.3 Spectral properties of land surfaces

The land surface is a highly complex boundary, and so it follows that the spectral properties of that boundary are considerably harder to define than the relatively simple cases of the ocean surface or the atmosphere. The Earth's surface can be rocky, sandy, dusty, muddy, swamp-like, covered in grass, shrubs, trees, in any combination. The analysis of the spectral properties therefore becomes a study of the

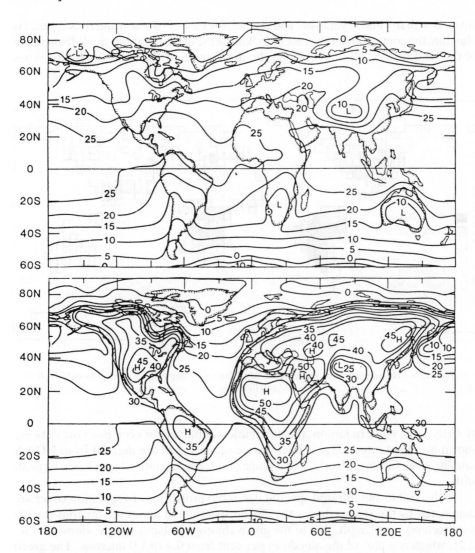

Fig. 2.19 — Influence of soil moisture on surface temperatures: the upper curve shows a model simulation for a saturated (wet) land surface; the lower curve shows the same but for a dry land surface. The moderating influence of moisture on land surface temperatures is evident.

various types of cover or material which we can find at the surface. The problem has been extensively treated elsewhere, for example see Kondratyev (1969) and Ulaby, Moore and Fung (Vol. III, 1986), and so we will consider only a few fairly simple examples for the sake of illustration here. The reader should bear in mind that because the land surface is so highly variable, there has to be a good deal of empiricism in the measurement and description of the spectral properties, more so than in the atmospheric or oceanic cases.

Two broad categories of land surface are of interest. First the case of land surface covered with soil and vegetation; and second, the case of bare soil and rock types. To illustrate the complexity of the task of making sense of observations made from space of the land surface, Fig. 2.20 shows the distribution of the main vegetative types

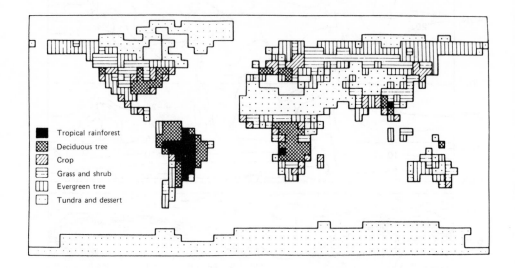

Fig. 2.20 — Distribution of major ecotypes on a 5 degree × 5 degree grid, derived from NOAA satellite data.

globally, or in modern jargon, the distribution of the 'major ecotypes'; the data are shown averaged into 5 degree by 5 degree squares, and are derived from NOAA satellite observations. Clearly, a vast range of surface vegetation types has to be dealt with.

In the visible and infrared spectral bands the properties of vegetation cover are dominated by the sharp edge in the albedo spectrum at about 0.7 microns (= 700 nanometers) wavelength, due to the green chlorophyll band. This is shown in Fig. 2.21, which is a plot of the albedo in per cent from 0.4 to 1.0 microns. The green colour of vegetation is caused by the rapid rise in albedo (reflectivity) in the green portion of the spectrum, which continues well into the infrared. The Figure shows some data for bare soils (both wet and dry) for comparison. It is possible to use this rapid change of spectral properties with wavelength to place measurements from space on a more quantitative footing. We can define a parameter called the Normalized Difference Vegetation Index, NDVI, as

$$\text{NDVI} = \frac{R(\text{NIR}) - R(\text{VIS})}{R(\text{NIR}) + R(\text{VIS})}, \tag{2.57}$$

or the difference between the reflectivity in the near-infrared, $R(\text{NIR})$, and the reflectivity in the visible, $R(\text{VIS})$, normalized by the sum of the two values. The NDVI has been used to quantitatively relate the radiances measured from space to

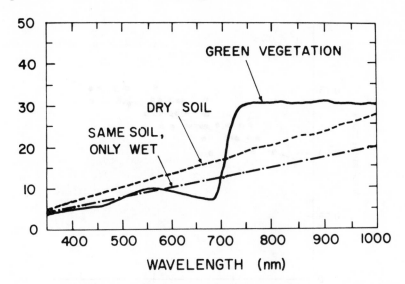

Fig. 2.21 — The 'green edge' in the reflectance spectrum of vegetation, compared with the reflectance of dry and wet soil.

the actual amount of live vegetation in the field of view of the space sensor, with a considerable degree of success. For example, Fig. 2.22 shows data obtained by the Landsat spacecraft, which indicates a reasonably useful linear relationship between what is called 'live biomass' (in g m^{-2} of 'dry weight') and the NDVI. We will see later other examples of quantitative estimates of surface cover from satellite observations.

Also, in the visible and near-infrared, minerals found in rocks and soils exhibit a number of clear characteristic signatures in their reflectance spectra, which in principle allow the geologist to make identifications of mineral types at the surface from cloud-free visible reflectance spectra taken during daylight. These minerals also often have strong absorption/emission features in the mid-infrared, which may also be used by day or night. To illustrate this, Fig. 2.23 shows the reflectance spectra of a number of minerals, containing carbonate, sulphate, and hydroxyl ions, or transition elements such as iron, Fe, titanium, Ti, or manganese, Mn. The markers on the various curves show the spectral channels that have been chosen for a new space sensor called the High Resolution Infrared Spectrometer, HIRIS, that is planned to be flown on the US polar platform in the late 1990's (NASA (1988b), EOS Vol II).

In the microwave region, consideration of the spectral properties of soil is complicated by the same aspects as in the visible, though we find that the microwave emissivity of the surface is dominated by two effects: the roughness of the surface in comparison with the wavelength of the radiation used; and the moisture content of the soil or vegetation. We also find that effects such as the Brewster angle effect that we found occurred in the reflectance of sea water also occur at the land surface especially where bare soils are not roughened by the presence of vegetation. Fig. 2.24 shows some calculations of the 'brightness temperature' of an homogenous soil surface as a function of angle of incidence, and for horizontal and vertical polariza-

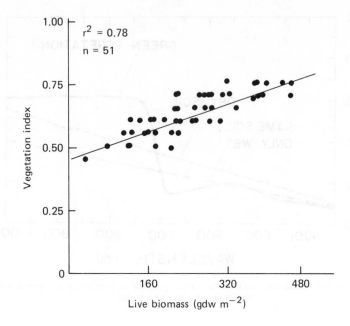

Fig. 2.22 — Live biomasss versus the NDVI (see text) based on Landsat data.

tions. The brightness temperature, T_B, is simply the true temperature multiplied by the surface emissivity

$$T_B(\lambda) = \varepsilon_\lambda \cdot T \,. \tag{2.58}$$

The true surface temperature in these calculations was taken to be 300 K, and the three cases included 0.05, 0.20, and 0.35 g cm^{-3} of moisture, respectively. The figure also shows the values of the complex dielectric constant that were derived in this study.

In recent years, it has been discovered that the polarization of the microwave radiation emitted by a vegetation-covered surface is a very useful parameter when it comes to trying to determine information about the amount of vegetation present at the surface. This may be understood at a simple level on the basis of the increased roughness of a vegetation-covered surface 'scrambling' the difference between the emissivity at the two different polarizations. In more technical terms, the vegetation induces a strong de-polarization of the microwave radiation. For example, the difference between the vertically and horizontally polarized brightness temperature at 37 GHz frequency is about 25 K over bare soils, but is reduced to about 5 K over short crops (e.g. alfalfa). The microwave polarization difference temperature, MPDT, is given by

$$\text{MPDT} = T(\lambda,\text{hor}) - T(\lambda,\text{vert}) \tag{2.59}$$

where 'hor' and 'vert' refer to the temperatures at wavelength λ, measured in horizontal and in vertical polarization respectively.

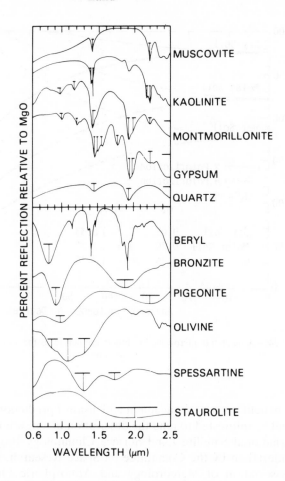

Fig. 2.23 — Visible and near-infrared spectra of minerals.

Thus the polarization difference at 37 GHz can be used to remotely sense the biomass at the surface, as is illustrated in Fig. 2.25, which shows the polarization difference for three locations, the Sahara, the Senegalese Sahel, and West Texas/ Oklahoma. The annual 'green wave' of growth in the Sahel can be seen as evidenced by the sharp decrease in polarization difference as vegetation blooms. The data also show evidence for the ecological disaster which has occurred in the Sahel since 1983 when a major drought occurred: it can be seen that the annual vegetation-induced minimum in polarization difference is weakened, and the conditions in the Sahel start to approach those in the Sahara in 1984 and 1985. We have seen all too starkly the results of these changes as they affect the people who live there, as on TV we see people starving, and major relief efforts being necessary to help them.

Making quantitative sense of observations of the surface from space is difficult, because of the complexity of surface types, and because of the obscuring effects of clouds. Nevertheless, this is a vital area of climate remote sensing which is potentially

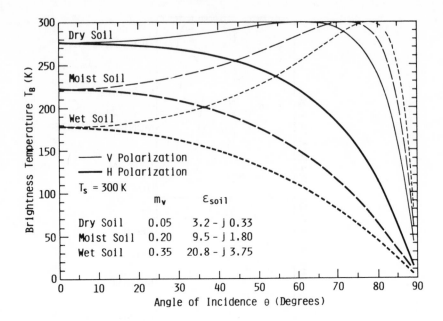

Fig. 2.24 — Calculated microwave brightness temperatures for a homogeneous soil at three levels of moisture.

of direct benefit in crop management and disaster prediction and management. The reader will be interested to read the description of a new international programme, the International Satellite Land Surface Climatology Project, ISLSCP, which is being undertaken by the Committee on Space Research, COSPAR, the International Association of Meteorology and Atmospheric Physics, IAMAP, in co-operation with the World Climate Research Programme. This description can be found in a useful little booklet titled *The International Satellite Land Surface Climatology Project* by Becker, Bolle and Rowntree (Becker *et al.*, 1987).

2.5.4 Chemical effects, dust, and the biosphere

We have so far discussed the thermal and radiation balance properties of the land surface. There are, however, other less direct influences of land on climate, and vice versa, which nevertheless may have a major role to play in our global climate.

Firstly, the land surface can act as both a producer (source) of atmospheric gases, and an absorber (sink) of these chemicals. Thus marshes, animal and vegetation waste and decay all produce methane (CH_4 — marsh gas), which plays a role in the water cycle in the upper atmosphere (where CH_4 is oxidized by other trace gases such as atomic oxygen, O, to produce water vapour, H_2O). Methane is also important in the global greenhouse effect, as an absorber of infrared radiation in its own right, as we shall see in the next chapter. Conversely, the land surface and particularly vegetation cover can take up chemicals from the atmosphere, and act as a moderator of the concentration of these gases in the atmosphere. For example, forest canopies

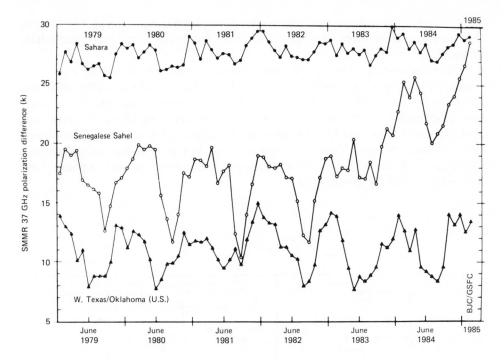

Fig. 2.25 — Difference between vertically and horizontally polarized microwave brightness
temperature for North Africa and for the USA.

can absorb significant quantities of CO_2, and thus play a role in reducing the atmospheric greenhouse effect. Removal of such canopies, as is happening at an alarming rate as a result of tropical deforestation by man, can enhance global warming.

A second aspect of the influence of land surfaces on the atmosphere and the climate is through the production of dust and aerosol particles. The most extreme example is the volcanic eruption, which often deposits huge quantities of material into the upper air, sometimes with very marked and directly linked consequences, such as a significant cooling of the stratosphere as happened after the Mt. St. Helens eruption in 1980. However, more insidious and long-term effects are thought to arise from over-cultivation and consequent drying-out of soils, thus producing more airborne dust. This problem is obviously important in many of our current concerns about the climate, though the precise effect of such dust is far from clear. What has been shown by work to date, for example, is that air laden with Saharan dust over the Atlantic could increase heating rates in the lower troposphere (from the surface to perhaps 500 mb) significantly (by about 1°C per day), and thereby cause a southward shift in the inter-tropical convergence zone (ITCZ). This could further reduce any chance of precipitation over the Sahara and Sahel regions. Other processes are, however, thought to be important in desertification, such as soil moisture and evaporation, albedo changes, ocean-to-land energy gradients, and the CO_2 effect, and our understanding is presently at best crude. Given the impact on people living in these regions, it behoves us to give such problems greater and more immediate attention.

The recognition that living organisms, plants and animals (including, of course, humans) play a direct, interactive role in the climate has brought about the coining of a new term, the 'biosphere', to capture all these disparate concepts into a single handy category. Present-day efforts at modelling such processes are at an early stage of development, and considerably more work is required. Bolin (1984), has developed a model, which is shown in schematic form in Fig. 2.26, for the storage and exchange of elemental carbon between various components of the biosphere (land and oceanic) and the atmosphere.

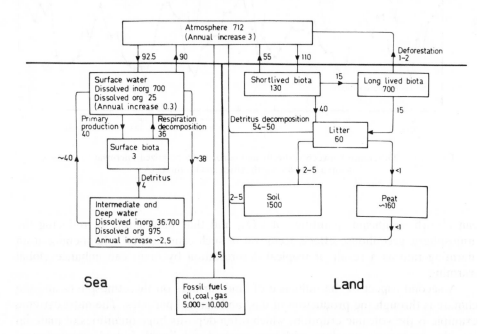

Fig. 2.26 — Major features of the carbon cycle. Masses in boxes are given in units of 10^{15} g; fluxes between boxes are in units of 10^{15} g yr^{-1}.

The advent of the concept of the biosphere has also led to the establishment of new international efforts, most notably the International Geosphere–Biosphere Programme, IGBP, to coordinate research efforts to study the complete, integrated system of 'Planet Earth'. NASA has even developed the concept of the 'Mission to Planet Earth', which may be something of an emotive title, but which serves the very useful purpose of drawing both attention and funds to the vital task of understanding how our own planet works. A description of the initial priorities for an IGBP can be found in a booklet published by the National Academy Press, Washington DC, in 1986, under the title *Global Change in the Geosphere–Biosphere*.

2.6 THE CRYOSPHERE — ICE AND SNOW

The 'cryosphere' is a term that has been coined to include the regions of the Earth where processes are effectively operating at temperatures usually below the melting point of water. The cryosphere is generally taken to include:

— seasonal snow
— sea ice
— the great ice sheets (e.g. Antarctica, Greenland)
— mountain glaciers
— permafrost

The vital statistics of the cryosphere are given in Table 2.1, from which a comparison shows that the relative volumes of the world's oceans and the world's ice masses are as follows:

$$\text{global volumes: } V_i = 31 \text{ million km}^3,$$
$$V_0 = 1370 \text{ million km}^3.$$

The Table also illustrates the great difference in depth between the Arctic, which composes mainly floating sea ice, and the great ice sheets of Antarctica and Greenland which comprise grounded sheets with thicknesses of several kilometres.

2.6.1 Water storage and sea level

Since (Table 2.8) an estimate of the effective mean area (A_o) of the global ocean is 361 million km^2, we can make a simple estimate of the change in ocean mean depth (= sea level) due to a given change in ice volume, for example due to a global warming. It is emphasized that this simple calculation is done solely for demonstration purposes, and ignores many complicating (but important) factors, such as the time constants involved in melting the large ice masses (this would be an extremely long process), and related feedback processes such as increased precipitation over the ice sheets, owing to enhanced atmospheric humidity, also brought on by the same global warming. Nevertheless, it is interesting to go through the calculation of sea level rise, if only so that the reader can begin to experiment with his own estimates of disaster scenarios as they have become known in the USA!

Returning to our simple calculation, if we assume that a fraction, f, of the global ice inventory is melted, then the change in depth of the oceans, Δx, is given by

$$\text{Volume change} = A_o \times \Delta x = f \times V_i . \qquad ((2.59)$$

Table 2.9 shows values of Δx, the change in global sea level, for a given fractional melting of the great ice masses. From this table we can see that the problem is not an

Table 2.9 — Change in sea level, Δx, for a fractional melting, f, of global ice sheets

f	Δx (m)
0.01	0.86
0.03	2.60
0.10	8.60
0.30	25.60
1.00	85.50

$\Delta x = f \times V_{ice}/A_o$.

insignificant one : a complete melting of the ice (ignoring other effects, as we noted above) would cause a rise in sea level of almost 90 m. Bearing in mind that most countries have economically and socially important cities at the edge of the sea, and that a rise of a few metres in many cases would cause severe flooding, we can see that even a small fractional melting of the global ice volume could have serious consequences. As we shall see later, forecast warmings in polar regions of up to 8°C could give rise to a large rise in sea level, though we should also note in passing here that direct thermal expansion of the oceans as a result of global warming might give rise to larger and more immediate sea level rises: also it is likely that though they contain much smaller stores of water, the mountain glaciers and small ice sheets might produce a more rapid response to global warming than the large ice masses, simply because they posess smaller thermal response times owing to their smaller sizes.

According to Untersteiner (1984), the characteristic times for interactions of the various elements of the cryosphere with other components of the climate system depend on the nature of the processes involved, and can be given by:

- Ice sheets and permafrost: 10^3–10^5 years
- Temperate glaciers: 10–10^3 years
- Sea ice and snow: 10^{-2} – 10 years.

It has to be remembered, therefore, that the immediate climatic impact of, say, a change in atmospheric temperature will not depend on the processes within the deep ice sheets, but on the processes relating to sea ice and snow melt at the sea ice boundaries and the surface, and those relating to glaciers.

Our understanding of the mechanics of ice sheet melting is far from complete: the mechanical properties of ice sheets are known to be strongly dependent upon the interaction at the interface with the ground at the bottom boundary. Sudden, catastrophic releases of ice into the oceans are known to occur, but the precise nature of the mechanisms which can trigger such surges is not understood. There is, therefore, a possibility of a catastrophic event occurring if a large part of the Antarctic ice sheet, for example, suddenly 'calved' into the circum-Antarctic ocean. Such processes would occur on top of the large annual variability in sea ice and snow presently known to occur (e.g. see Fig. 2.27).

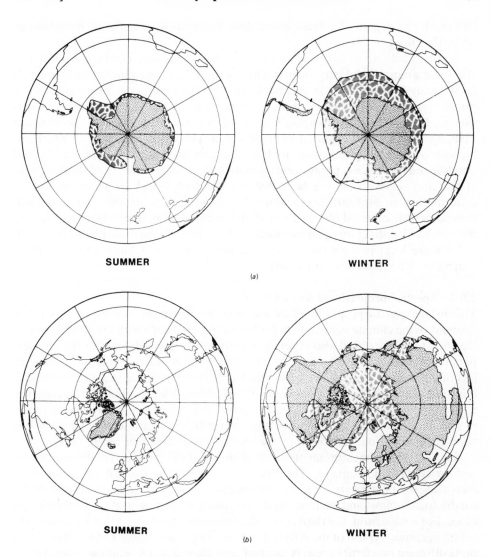

Fig. 2.27 — The annual variability of sea ice around Antarctica (a), and in the Arctic (b), for both summer and winter.

2.6.2 Radiative exchange effects
The influence of ice and snow in the complex radiative feedback of the climate system is also important. This can be illustrated by one simple example. Imagine that global temperature rises lead to a reduction in areal ice extent (more open water). This will reduce the reflection of solar radiation, and lead to an increase of absorption of solar energy by the upper layers of the ocean. This would lead to a further rise in temperatures, i.e. a positive feedback effect. However, we might also expect increased humidity, and therefore (possibly) increased cloudiness. This would in

turn produce a negative feedback effect, since incoming solar radiation would be cut off by the cloud.

Similar qualitative examples can provide interesting excercises for the mind, but also serve a more useful purpose in reminding us of the theme that has been touched on so many times already in this book, that is, the enormous complexity of the climate system, and of the responses we must expect of that system to any imposed change. In studying the cryosphere, as in other parts of the climate, the ultimate challenge is to develop quantitative models of the processes in the system, which are capable of accurate predictions of how the various competing, interfering and simultaneous processes operate, and ultimately of accurate forecasts of the final state of the climate. We shall see later that the development of an understanding of cryospheric processes sufficiently accurate to allow us to model the system well will depend heavily on good observations of global parameters from space (e.g. sea ice distribution, ice sheet topography, ocean and ice surface temperatures, etc.), as well as on more local process studies (e.g. melting, freezing, salinity effects, surface roughness, ice mechanics, and so on).

2.6.3 Spectral properties of ice and snow

The surface spectral properties of ice and snow are important, as is the case for other domains of the climate system, both for the energy exchange with the atmosphere in the climate balance, and also in terms of methods of remotely sensing the physical parameters of the surface from space. The spectral properties are complicated by the fact that the surface can be in a very cold, highly frozen state, or in a warmer slushy, mixed ice, snow and liquid water combination. Also, at least in the microwaves, the scale of the surface roughness of ice and snow, in relation to the wavelength of the radiation, affects the spectral properties. Nevertheless, considerable progress has been made in our understanding of the spectral properties over recent years.

Fig. 2.28 shows measurements of albedo made at the surface over three different surface types in Greenland. The two curves which show a fall in albedo with increasing wavelength are for snow, the upper curve (higher values of albedo) being for dry frozen snow, and the lower of the two being for wet snow at the melting point of ice. For comparison, the third curve, shown increasing in albedo with wavelength, is for vegetation typical of the Arctic, for example grasses, mosses and willows. The quite different properties are very marked, and allow easy recognition of vegetation or snow. The difference between dry and mushy snow, however, is not so clear, particularly when it is remembered that any typical satellite sensor scene is likely to include rather a range of surface conditions. Nevertheless, these data do show a way of making climatically useful measurements of the surface from space.

The detailed treatment of the spectral properties of snow and ice in the microwave region can become very involved, since the properties depend strongly on the exact state of wetness, and on the details of the surface roughness. However, again, considerable work has been done, and many results may be found in the literature. A good reference is *Microwave Remote Sensing* by Ulaby *et al.* (1986), from which we have taken Fig. 2.29. This shows the brightness temperature at both horizontal and vertical polarizations of three types of surface, wet spring snow, dry winter snow, and refrozen spring snow. The spectral properties are evidently very different. The major differences are attributed to the differences in the shapes and

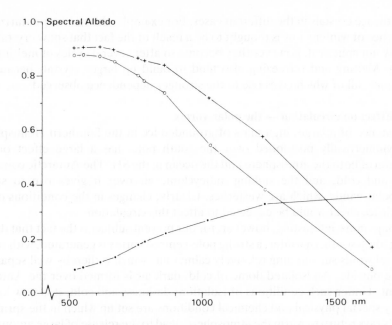

Fig. 2.28 — Observed spectral properties of three types of surface in Greenland: (+) dry, frozen snow; (O) wet snow at melting point; (.) arctic grass, moss and willow.

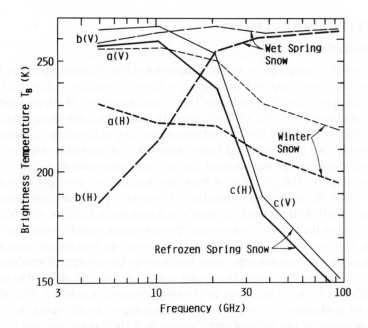

Fig. 2.29 — Microwave spectral brightness temperatures for both horizontal and vertical polarizations, for wet spring snow, dry winter snow, and for refrozen spring snow.

sizes of the ice crystals in the different cases. For example, the stronger polarization dependence of winter snow is thought to be a result of the fact that snow crystals are generally not spherical: however they become so after several cycles of melting and freezing. Melting and refreezing also tend to build up larger crystals because of coalescence, all of which gives rise to the stronger dependence observed.

2.6.4   Effect on circulation — the polar vortex

The existence of a large, high mass of grounded ice in the Southern Hemisphere, fairly symmetrically positioned over the south pole, has a large effect on the circulation of both the atmosphere and the ocean in the SH. The Antarctic continent is high and cold, and the sinking anticyclonic air over it gives rise to strong circumpolar winds, which are westerlies. Clearly, changes in the conditions of the Antarctic ice sheet would be expected to affect this circulation.

Perhaps more interesting, however, for our present subject is the fact that during the long, dark southern winter, a strong pole-centred vortex is generated, with strong peripheral winds surrounding relatively calmer air, which is thereby well separated from the outside. An isolated dome of cold, dark air is formed over the Antarctic continent. Under such conditions of isolation, darkness and cold, it is now known that very special physical and chemical conditions are set up which in the spring, as the Sun's rays return to warm the atmosphere, lead to the release of large amounts of free chlorine which can destroy ozone and form the now-notorious 'ozone hole'. More on this in Chapter 3.

2.7   MODELLING THE CLIMATE

2.7.1   Introduction

We have described some aspects of the complexity of the climate system in what has already been discussed, and we have considered a number of the radiative, chemical and dynamical processes that make up this complex system. We have, naturally, concentrated somewhat on the radiative aspects, simply because these are the ones that most concern us in this book about remote sensing of the climate system from space. However, we have provided some examples to demonstrate that important chemical and dynamical processes must also be taken into account.

We now come to the question of how the theoretician attempts to study the effects of all of these processes together, occurring simultaneously, and often interacting strongly with each other, usually in a non-linear fashion, with feedback. The answer is that the theoretician constructs a computer-based 'model', or simulation, of the climate system, representing each important process as accurately as possible through a set of equations, some exact, some known approximations to the truth. Once properly formulated, the model can be started, and allowed to run until some equilibrium between the many interacting processes is reached. The accuracy of the model is steadily improved until it is capable of reproducing at least the important features of our present-day climate; it is then ready for use to predict future trends or the past climate record if we introduce likely or possible perturbations into the formulation of the model.

This all sounds beautifully simple, but of course this is very much an idealized description. The real climate system is vast, and operates at a range of scales from thousands of kilometres down to the minute, from millenia to minutes, and so the representation in a finite computer program is an extremely difficult task. In this section we will give an introduction to some of the progress that has been made in modelling the climate system.

One further point. It could be said (indeed I have heard it said by at least one theoretician!) that by far the most important activity to fund in climate research is modelling, even to the exclusion of all else. This argument ignores the fact, of course, that our models are always going to be imperfect representations of the climate, simply because of our imperfect knowledge of the processes involved, as well as our inability to perfectly simulate those processes in a computer (if only because of the limitations of computer size). It would be more reasonable to accept that the development of models, and the improvement of our capability to observe the global climate should go on hand in hand, since both depend on each other. We use models to predict the future trends in climate, but we also need accurate observations in order to test the accuracy of those models, and also to watch out for changes in the climate system which we have failed to predict with our models — the appearance of the Antarctic ozone hole being one such example.

This section provides a brief introduction to modelling of the climate system, it is to be hoped at a level sufficient to provide the reader with some insight into the scientific principles being used, and the current status of what is a very difficult task, i.e. an accurate simulation of the global climate in all its complexity. We will not attempt a critical review of the technical mathematical techniques which are used in these models: such details are the province of more advanced texts than this one. Thus, the reader is referred elsewhere for more thorough treatments, particularly to the recent report of Working Group 1 of the Inter-governmental Panel on Climate Change (IPCC; — see Houghton (1990)), and other references including, Manabe and Bryan (1969), Manabe and Stouffer (1988), Mitchell *et al.* (1987), Hoskins *et al.* (1983) and Gates *et al.* (1985).

2.7.2   What is a model? The case of the Atmospheric Global Circulation Model (AGCM)

The beginnings of modelling studies of the climate are to be found in meteorology. Soon after the invention of the digital computer, meterologists recognized the possibilites for simulating the many different processes that were going on in the atmosphere, and thereby creating a description of the weather system that would permit forecasts of the future state of the atmosphere (an interesting account of the early days of numerical weather prediction may be found in Smagorinsky (1983)).

The construction of a model depends on the degree of complexity to be attempted. There are many ways that have been developed to simplify the task, tailored to specific problems, each with its own idiosyncracies. It is not appropriate in a book of this nature to discuss all such types, and so we will concentrate on describing the broad features of a general circulation model (GCM), working in three dimensions of space, and of course one dimension of time. (At the end of this section we will briefly outline some of the other types of model that have been used in recent research efforts, to give the reader some further background.)

The basis of the model is a set of equations (the so-called 'primitive' equations), which set out the basic dynamical, thermodynamic and conservation properties of the atmosphere in a quantitative way; the conservation equations ensure that certain fundamental properties such as energy, mass and momentum cannot mysteriously appear or disappear in the model, but must be conserved in one form or another as the model does its business. The parameters appearing in these equations include the temperature, the pressure, the wind components, the moisture (for an atmospheric model), and the density, currents and salinity in an oceanic model. One complexity that must be dealt with is that in the model these equations must operate on a spherical Earth, and so the equations have to be mapped into spherical coordinates. We can follow Gill (1982) and others in setting out, in simplified form, some of these basic equations here (ignoring friction):

$$\text{Momentum (incompressible fluid): } \frac{D\mathbf{v}}{Dt} - f\mathbf{v}\wedge\mathbf{k} + \frac{1}{\rho}\nabla P + \mathbf{g} = 0 , \qquad (2.60)$$

$$\text{Thermodynamics: } \frac{DT}{Dt} - \frac{RT}{SP_0}\frac{DP}{Dt} = 0 , \qquad (2.61)$$

$$\text{Moisture conservation: } \frac{Dq}{Dt} = 0 , \qquad (2.62)$$

with other similar equations for mass conservation, hydrostatic equilibrium, and so on. In these equations, f is the Coriolis parameter (see equation (2.41)), $\mathbf{v}$ is velocity, $\mathbf{k}$ is a unit vector in the vertical direction, P is pressure and ρ is density, $\mathbf{g}$ is the acceleration due to gravity, T is temperature, S is the specific heat of dry air at constant pressure, and q is the specific humidity. (The reader will recognize the similarity between the momentum equation depicted here and the equations of motion describing the dynamics of atmospheric and oceanic fluids given earlier in this chapter; see equations (2.41) and (2.53) for example.)

In these equations, the symbol D represents a partial derivative of a given quantity for a particle of air which is in motion (see Gill, 1982, p. 64). Also, in the model, each of these equations is equal to zero only if there are no external processes which perturb the balance of momentum, heat, moisture, and so on. In practice, terms have to be inserted on the right-hand side of the equations to allow for the fact that processes of radiation, convection, turbulent mixing and precipitation can introduce perturbations to these, and the other, balance equations. The model may also need to include on the right-hand side of these equations a term to ensure that the scales of motion which occur on dimensions smaller than the resolution of the model are adequately treated, and that any smoothing of the data fields in the model does not give rise to unrealistic perturbations.

The model can be set up on the basis of a 3-D array of grid points around the globe, in which case (e.g. see Hansen et al., 1988) a typical model might have up to about ten layers in the vertical, and a horizontal separation between grid points of perhaps 5 to 10 degrees of latitude and longitude; currently some research centres are developing 'high resolution' models with grid point separations of 2.5 degrees globally, which require the biggest, fastest computers available for their use. An alternative approach is to use a spectral transform technique in which a parameter such as pressure is described in terms of a series expansion of spherical harmonics around the globe (see Hoskins and Simmons, 1975). It has been found that in the latter case, about 15 terms in the series expansion are required to achieve a reasonable level of realism, and again up to 10 or so layers in the vertical has been typical.

We have already considered in part the processes which must be included in a realistic AGCM, but it might be useful to collect some of them together here:

— heating (e.g. by radiation)
— cooling
— evaporation
— condensation
— liquid water processes
— cloud formation
— cloud-radiation interaction
— sub-grid-scale parameterizations.

Of these processes, it is perhaps the last two that remain major problems. The feedback of cloud and radiation is a desperately complex problem, as we saw in section 2.3.1, and the parameterization of this in models needs to take account of all scales from the near-global to the very localized. Considerable work, worldwide, is currently being devoted to the cloud problem. More generally, the 'sub-grid-scale parameterization' problem, of how to represent in an equation all the processes which may be occurring at scales below the resolution of the model is extremely challenging; usually it is necessary to develop the necessary equations empirically from observations. For example, this is one of the goals, as far as the cloud problem is concerned, of the International Cloud Climatology Programme (ICCP; see World Climate Programme Report WCP-95, 1985). Often, modellers find that it is necessary to 'tune' these parameterizations in order to get the model to produce the 'right' behaviour, in which case one can never be sure if the right behaviour is occurring for the right physical reasons or not!

Also of importance is the achievement of an accurate description in the model of the important feedback processes. We saw earlier how important these are for a thorough understanding of the climate. Feedback mechanisms can, of course, either amplify or reduce the response of the climate system to an induced change (a climate 'forcing' in the emerging jargon of this subject). Well-known feedback processes include the effect of changing cloudiness consequent on an increase in global mean temperature, which can either amplify the warming if the increased cloud cover produces an additional warming, or decrease it if the effect of the cloud is a net cooling. Another example of a feedback process is the increased atmospheric humidity expected as a result of global warming. Water vapour itself is a strong natural greenhouse gas, and an amplification is possible. The feedback of changes in the snow and ice cover is another example: if sea ice coverage decreases due to global warming, then the net absorption of incoming solar radiation will increase, which will enhance the warming (unless, of course, the increased atmospheric humidity leads to increased cloudiness, which cuts off more of the sun!). As we have seen before, these highly interactive processes can really only be accurately simulated in a computer model which allows them to develop and interact simultaneously much as they do in the real world. The accurate expression of these feedbacks in a model is therefore of great importance.

So far, we have gone some way to describe the make-up of an atmospheric model (AGCM). However, a purely atmospheric model is not a climate model; a climate model has to take into account the substantial effects of the oceans, the land and the cryosphere. Many atmospheric models prescribe fixed sea surface temperature (SST) in order to provide a lower boundary condition for the purposes of calculation. For a model to be called a true climate model, it must include an interactive ocean, and a realistic representation of the land, the cryosphere, and the most important biogeochemical processes.

2.7.3 Ocean, cryosphere, land surface and biosphere models

The oceans represent, as we have seen, a very large thermal 'reservoir' in the climate system (Ramanathan, 1981). As a result of the greater fluid density of water compared with air, we find that the oceans have a large thermal time constant compared to the atmosphere, ranging from weeks to many hundreds of years. Similarly, the space scales typical in a denser fluid are shorter. Though we might expect the longer time constant to require a longer time step in the computational process, and therefore less computer speed, the smaller spatial scales mean that a much finer grid must be used than in the atmosphere. Taken together, the increased demands from the spatial problem outweigh the reduced ones from the time constant, and generally ocean general circulation models (OGCMs) represent a more difficult computational problem than AGCMs.

The parameterization of sub-grid-scale processes remains a major problem in ocean modelling, as it does for the atmosphere. Indeed, this is probably a more serious difficulty, because of the greater sparsity of measurements which are available for the ocean, making the development of reliable algorithms for describing such processes more difficult. Largely because of a recognition of this relative shortage of global data for the oceans, world programmes have been set up in the past decade to provide the measurements. Examples of these are the World Ocean Circulation Experiment (WOCE), and the Tropical Ocean–Global Atmosphere (TOGA) programmes, set up under the auspices of the World Climate Research Programme. (Anyone interested in reading more about these programmes might refer to Woods (1983).)

The other major area of development required in OGCMs is an improvement in the description of the coupling of the ocean surface with the atmosphere, which through wind stress exerts a driving influence on the surface layers of the ocean, and which is also of importance in the exchange of energy (e.g. sensible and latent heat), momentum and moisture. Leith (1983) describes attempts which have been made to achieve 'asynchronous' coupling, by developing atmospheric models which produce statistical time-averaged states of the atmosphere, which are fixed and then allowed to interact with the ocean until a new surface temperature balance is set up: the statistical model is then run again to reach a new time-averaged state, and the process is repeated. Success has been limited, according to Leith, the basic incompatibility of time and space scales of the ocean and atmosphere presenting us with a fairly fundamental barrier to any short cut solutions. It is difficult, therefore, to avoid the need to develop a major coupled model, with space and time scales and steps adequate for both the oceans and the atmosphere.

Turning to the cryosphere and the land, the situation for modelling of these components of the system is in principle simpler than for the oceans. We have considered some of the salient features of the role of these components in the climate system above, in sections 2.5 and 2.6. In these cases there is no motion analogous to ocean currents or atmospheric winds to take into account. The problem therefore reduces to a description of the exchanges of heat, moisture and chemicals between the surface and the atmosphere. The surface is fixed except insofar as the surface cover may change during the year due to changes of vegetation, snow cover and suchlike effects. The equations are in general simple, and the dominant parameters that must be taken into account include temperature, soil moisture, vegetative cover, ice and snow depth, sea ice extent and thickness, and so on. In many cases, for example evaporation or latent heat release, we have an exact knowledge of the equations governing the process. In other cases, for example in dealing with surface vegetation, or with the moisture content of the soil, the situation is just too complicated to rely on anything other than an empirical fit of radiative or other exchange processes to measurements which have been correlated with surface type, vegetation amount and so on. For this reason, the International Satellite Land Surface Climatology programme has been adopted within the World Climate Research Programme, with the objective of using global satellite observations, correlated with surface 'truth' measurements of vegetation and ecosystem types, to develop empirical relations between those surface types and climate parameters, which can be employed as parameterizations in models.

In what we have discussed so far, we have implicitly included processes within the so-called 'biosphere' in our treatment of the land and to some extent the oceans. Fig. 2.26 showed a schematic diagram of the carbon cycle, from the work of Bolin (1984). This illustrates the sort of processes which must be taken into account when including the biosphere in a climate model. Clearly there are important processes which occur on land and at the surface of the oceans which strictly must be taken into account in a full climate model; this has not yet been achieved. The physical processes, such as gas solubility and fractionation of carbon dioxide between air and sea, are well understood, but the biological processes at work (e.g. the take up of atmospheric gases by marine organisms) are not.

The feedback processes which must be included in such models of surface interactions with the atmosphere include, rather obviously, processes like wind stress, heat exchange, latent heat release, albedo feedback, evaporation, thickness of sea ice and its extent, vegetation type and cover, and biogeochemical interactions. Expressions for these many feedbacks, some positive, some negative, need to be prepared and introduced into the models. A recent example of work to elucidate the feedbacks between the atmosphere and the surface, including for sea ice, may be found in Ingram *et al.* (1989).

2.7.4 Coupled models

If we have developed the separate models to describe the atmosphere, the oceans, the cryosphere, the biosphere and the land, we can in principle couple these together, let the composite run, and see what we get. We have already identified one of the reasons why this is not so simple in practice — that is that the time and space scales that must be used for the oceans and for the atmosphere are very different, and the coupled model must either be capable of tracking processes at the finest resolution demanded by the smallest scales, or must find some way of reducing the huge computational load by introducing a simplification in one or more components of the coupled model. For example, one of the groups engaged on developing a large coupled GCM is that at NASA's Goddard Space Flight Centre, led by J. Hansen. In recent work (Hansen *et al.*, 1988) this group has combined a large AGCM with an ocean model that approximated some of the heat transfer processes acting in the surface layers of the oceans. For example, horizontal heat transport is fixed in the model (at values estimated for today's climate), and further approximations are introduced to account for energy exchange between the upper, mixed layer and the deeper ocean.

As well as this problem of incompatibility of time and space scales for atmosphere and ocean models, there exists a more technical problem which arises in these large coupled models. Because each component of the coupled model is no longer constrained or kept in check at its boundaries by the imposition of observed parameters (observed temperatures, wind stress, etc.), but is free to interact with the boundary presented by the 'adjacent' model, the coupled models tend to 'drift' away from a realistic state. To counter this, correction terms for the fluxes between the ocean and the atmosphere have to be introduced to neutralize the drifts.

To date only a limited number of groups in the world have been able to attempt to run a coupled climate model, because of the considerable resources of computers, money and people required. These groups have generally succeeded in coupling together an AGCM and a mixed layer only ocean model (using approximations such as the ones used by the Hansen group in order to account for the rest of the ocean). The procedure has been to develop a 'control' run based on present day conditions. The run is then repreated with changed conditions (in the jargon, with 'changed climate forcing') and the model run to equilibrium for these new conditions (e.g. the new conditions might be a doubling of the amount of CO_2 in the atmosphere). The difference between the control and the perturbed run is the response to the changed climate forcing. Integration times have typically been between 5 and 100 years. In some cases, as we will discuss elsewhere (e.g. see section 3.2.3), the control runs themselves have shown some remarkable and possibly highly instructive behaviour, such as unforced sudden switches in climate state, due presumably to internal noise in the model. We shall see more of the products of such experiments in Chapter 3.

It has been realized that the response of the climate to a gradually introduced change of climate forcing (for example, a slow build-up of CO_2), might actually be different from the predictions of a model in which a step function change of conditions is suddenly turned on. Attention is therefore being given to running transient, or time-dependent, experiments in order to better simulate the greenhouse effect of gradually increasing CO_2, for example (see Chapter 3).

2.7.5 Model results to date

The current status of climate modelling, according to the recent IPCC report (Houghton, 1990), is one of mixed success. Both AGCMs and OGCMs, and of course the coupled models, have considerable success in simulating the present, and in some cases the known past, climate, though not without worrying errors at times. For example, surface air temperatures can be wrong by up to 3°C, at least on the regional scale, and precipitation can be predicted only with an accuracy (depending on region) of 20 to 50 per cent. Other parameters such as snow cover and top of the atmosphere radiation balance are in some models well simulated, but not in others. Research has shown a strong sensitivity of results and errors to the resolution used in the models, and to the parameterizations used for sub-grid-scale processes, perhaps not surprisingly. Low frequency phenomena such as the El Nino (Chapter 3) are reproduced reasonably accurately. Also, runs of GCMs to simulate climates during the past 18 000 years or so have been broadly successful; for example, Manabe and Stouffer (1988) have shown that a coupled ocean–atmosphere GCM model shows signs of posessing two stable equilibrium states that may have relevance to studies of abrupt climate 'switches' that are thought to have occurred at times in the past.

It is clear that results to date are encouraging in the sense that improvements are being realized, and progress towards an ability to predict accurately is being made. However, it is equally clear that very considerable efforts are required in the development of the large coupled models with ever higher spatial resolutions, and in the parameterization of processes. Also, emphasis must be placed on acquiring further global data for the validation of the climate models. Thus we return to the primary theme of this book, which is, what can be measured globally from space? Before returning to our main theme, though, we will in passing review some of the types of models other than the GCMs that have concerned us here.

2.7.6 Other types of model

Not every research group has the resources to work with large 3-D GCMs, and indeed, there are many scientific investigations that are best carried out using other types of model. There has been a bewildering array of different model types used by scientists in their study of the climate, and in what follows we will mention just a few examples. These can vary from the full three-dimensional GCMs which we have just described, through models which reduce the number of independent dimensions to two or one, even to zero-dimensional models which treat the Earth as a single average point in space. Each of these model types has a particular use, a particular application for which they are ideally suited. For more details, the reader is referred to a more specialist text, e.g. Saltzman (1983), Monin (1986), Gill (1982); a non-specialist introduction is given by Schneider (1987).

Energy balance models

In these models the complexity of atmospheric and oceanic transport, and of radiative heating and cooling effects, is greatly simplified, in order to provide manageable computer programmes for examining interactive feedback processes, such as between clouds and radiation, or between temperature, ice cover and albedo (see Budyko, 1969). The simple calculation we did in section 2.1.5 represents the very simplest energy balance model that we can write down, a zero-dimensional case.

1-D and 2-D models

It is possible to use the quasi-zonal symmetry of the atmosphere (i.e. the property that conditions vary more rapidly with latitude than they typically do with longitude) to reduce the three spatial dimensions of an AGCM to two spatial dimensions (height and latitude), using weighted means of parameters to represent longitudinal effects. This reduces the computational requirements considerably, of course, while preserving the broad features of the atmospheric circulation. An even greater simplification is to reduce the dimensions to just one, i.e. height. Though all the horizontal variability of the atmosphere is lost, this can to some extent be recovered by making calculations for a variety of different locations.

The 1-D model is extremely useful for testing ideas about complex photo-chemistry, when hundreds of chemical reactions need to be included, but the expense and time of a full dynamical treatment can be put off. The 1-D model is also useful for radiative energy balance calculations: for example, a one-dimensional radiative-convective model of the atmosphere was used by Manabe and Wetherald (1967) to study the vertical distribution of global mean temperature in thermal equilibrium; such a model relies on a system of equations which represent the effects of radiative transfer and convective mixing on the heat balance of the atmosphere.

Stochastic models of the atmosphere
These are used to generate statistical information directly, rather than through time-averaging of a GCM, about a number of atmospheric parameters, such as surface stress, surface radiation fluxes, evaporation, humidity, clouds, etc. These are then used in conjunction with an OGCM, which, because it is varying more slowly, is reasonably compatible with the averaged atmospheric fields. From time to time, the averaged atmospheric fields can be recalculated, before the coupled model is taken further. There is still considerable research required to make the stochastic model fully representative of the true mean state of the atmosphere.

Noise studies
Investigations of the natural fluctuations, or 'noise', in complex GCMs have been carried out on a purely theoretical basis, by analysing the statistical correlation between developing time series of climate parameters, such as mean surface temperature or pressure. A correlation time, T_0, can be defined which is a measure of the predictability of the atmosphere (see section 2.1.8). Leith (1983) notes that the longer the correlation time, then the greater the range of deterministic weather predictability, but also the higher the weather noise level. Values of T_0 between 2 and 8 days have been deduced by various workers.

Linearized models
It is possible to remove the complex, time-consuming non-linear elements of a climate model under certain circumstances, to build a model that will reproduce the broad features of atmospheric or oceanic dynamics, for example, to study the way in which waves propagate. To do this, we may linearize the equations describing the flow of the atmosphere (for example), and this simplification preserves enough of the general features of atmospheric dynamics that phenomena such as wave channeling, absorption and dispersion can be demonstrated and studied. As an example, Gill (1980) used a linearized shallow water model of the equatorial ocean to show how

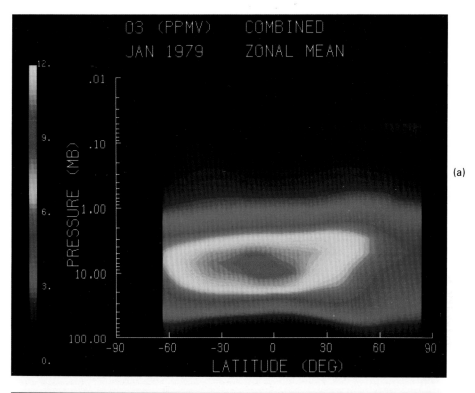

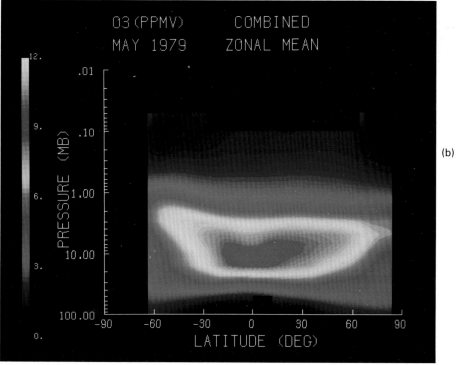

Plate 9 — LIMS measurements of zonally averaged ozone distributions for (a) January 1979, and (b) May 1979.

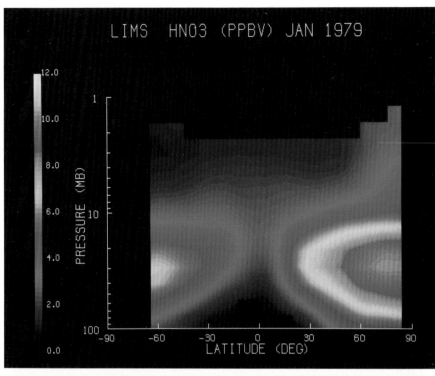

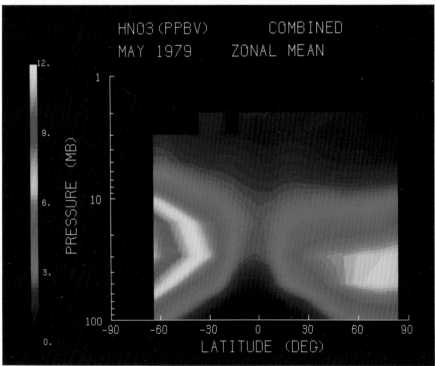

Plate 10 — LIMS measurements of zonally averaged nitric acid for (a) January 1979, and (b) May 1979.

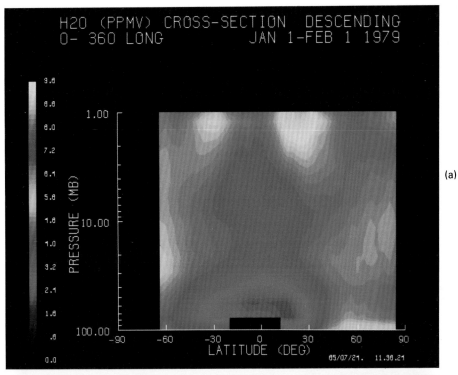

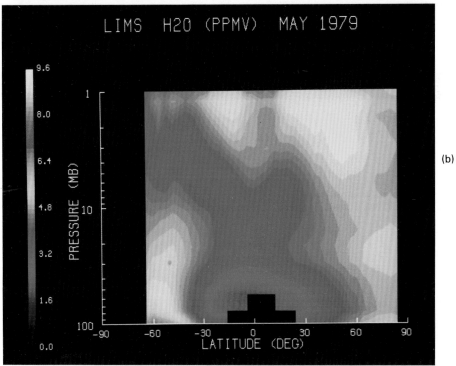

Plate 11 — LIMS measurements of zonally averaged water vapour for (a) January 1979, and (b) May 1979.

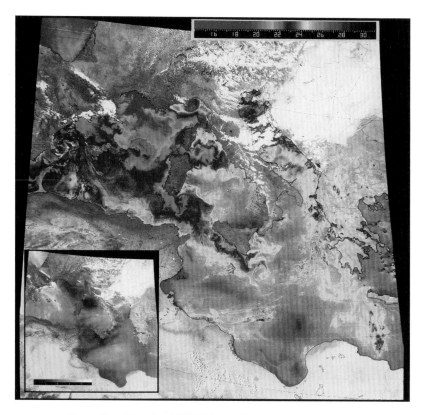

Plate 12 — An AVHRR image showing SST data.

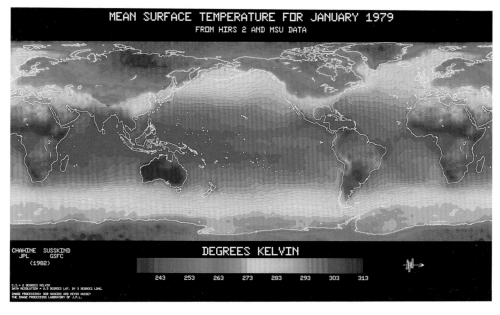

Plate 13 — Global sea surface temperatures derived from a combination of microwave and infrared satellite data.

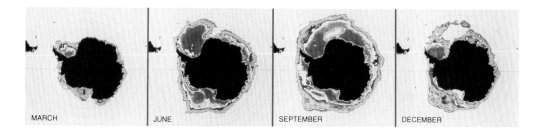

Plate 14 — Seasonal variations of Antarctic sea ice in 1974 from space.

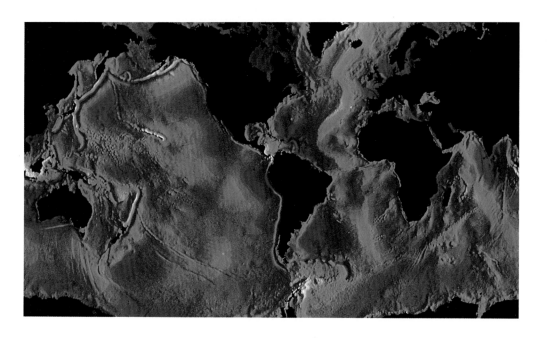

Plate 15 — Global ocean topography from SEASAT.

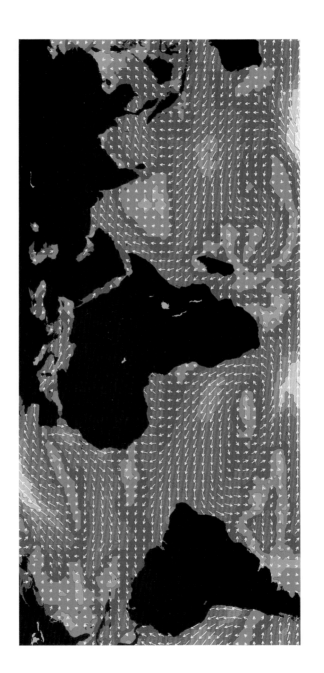

Plate 16 — Global surface wind field from SEASAT data.

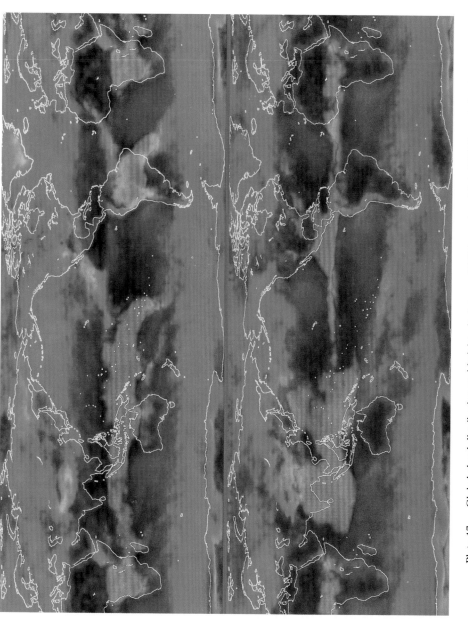

Plate 17 — Global cloud distribution and heights, in January 1979 (upper) and July 1979 (lower): low cloud, blue; mid cloud, green; high cloud, red.

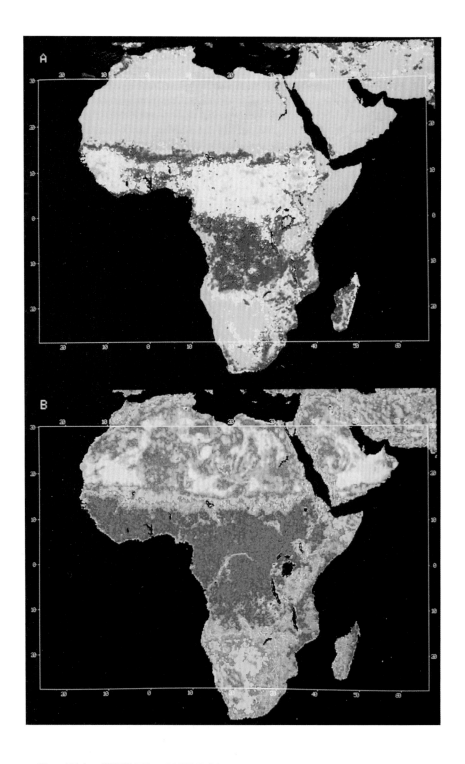

Plate 18(a) — NDVI (A) and MPDT (B) vegetation data for Africa: September 1982.

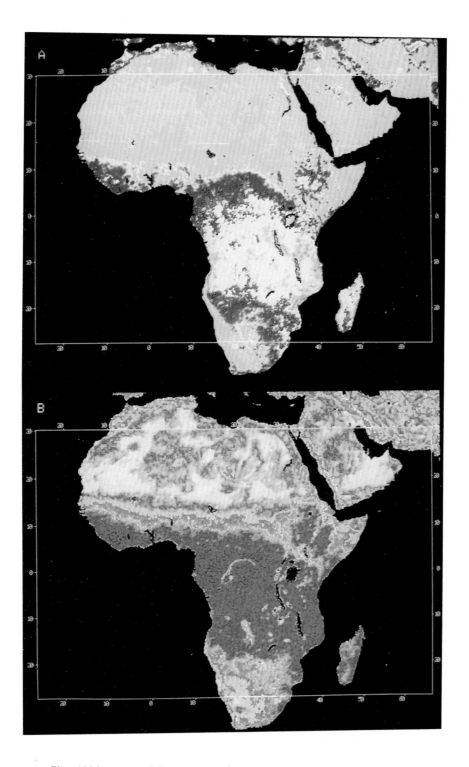

Plate 18(b) — NDVI (A) and MPDT (B) vegetation data for Africa: March 1983.

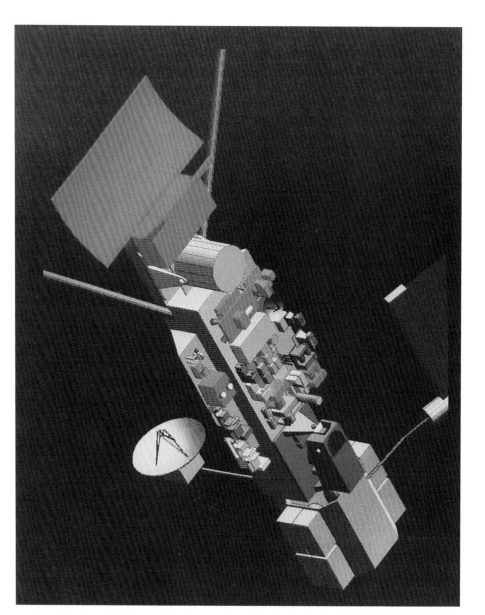

Plate 19 — An artist's impression of the US polar platform.

Kelvin waves are capable of propagating the effects of a localized source of heating in an eastern direction, in the way observed during an El Nino, thereby demonstrating some of the important features of this phenomenon.

Statistical–dynamical models

This is a class of model which simplifies the problem of massive computing by reducing the atmospheric dynamics to a zonal mean, or by limiting the resolution of the model to the longest longitudinal wavelengths only (several thousands of kilometers). The finer details of atmospheric behaviour are parameterized by averaging observations, or perhaps by averaging the results of a more detailed GCM. Because they are computationally cheaper to run than GCMs, these statistical–dynamical models can be used for studies of climate on much longer time scales than large GCMs.

This brief review of some aspects of climate modelling will provide the reader with background as we now go on in Chapter 3 to consider several specific climate problems in more detail, and then turn in Chapters 4,5 and 6 to consider the role of space observations in climate research.

3

Current climate issues: ozone holes, greenhouses, and El Nino

In this chapter we will be looking at some of the currently popular 'green' issues of the environment, especially from the perspective of space. Some of these issues have, in fact, concerned scientists for many years, and numerous learned papers have been written, for example, on the greenhouse effect. But we know that mere threats of climatic doom do not generally attract mass interest: far more captivating is the evidence of a climatic disaster actually upon us, and this is, of course, what has happened with the discovery of the Antarctic ozone hole. Suddenly there is irrefutable evidence of a major environmental change, which man was not capable of predicting in advance, which seems likely to have its roots in man-made pollution, and which could lead to climatic effects and even (and this is the clincher, of course), economic effects. Now environmental hazards are on the political agenda, and other important (though previously largely ignored) issues such as the greenhouse effect are also demanding attention.

The Earth can be regarded as a single system. Therefore, we should expect that different manifestations of environmental stress and change are linked to one another, more or less directly, and this is the case in the issues which form the title to this chapter. The polar ozone hole, the greenhouse effect and the El Nino all have links to each other, in some cases strong, in other cases weaker. Nevertheless, we shall be treating the three phenomena separately, and trying to distinguish the key processes which characterize them.

The first two of these topics reflect the influence of mankind upon the Earth, and at the heart of all such problems of anthropogenic disturbance, of course, lies the pressure caused by the growing numbers of human beings living their lives on the planet. Population growth leads to increased use of non-renewable (e.g. coal), or only slowly renewable (e.g. forest), resources, increased stress on land use, demands for ever greater productivity in food production, and greater burdens of waste products to be absorbed. Thus, the overriding problem is at heart that of over-population of the globe, a problem of immense political dimensions and great human sensitivity, the solution to which is consequently daunting in its scale and complexity.

The third of our topics, interestingly, concerns what appears to be a phenomenon of natural variability in the climate, caused by the interaction of the atmosphere with the vast thermal reservoir of the oceans, which we discussed earlier (see Chapter 2). Whether there are links to man's influence through processes like the greenhouse effect has yet to be determined.

3.1 THE OZONE HOLE

3.1.1 A primer on stratospheric ozone

What is the ozone hole? The ozone hole, a very recently discovered phenomenon (1985), is a depletion of the concentration of the global stratospheric ozone layer, which has been occurring locally over the Antarctic continent in the Austral spring, between September and November each year since about 1979.

Ozone is a gas, a chemical derivative of the oxygen molecules in the air we breathe. Instead of containing two oxygen atoms, as does molecular oxygen (O_2), ozone contains three atoms (O_3). It is a very reactive gas, even explosive in high concentrations, which is present in the troposphere in trace quantities. Its most important role is in the stratosphere, however. There, a dilute (few parts per million) layer of ozone extends from about 15 km altitude to about 40 km. This is really very rarified (the popular comment is that, if compressed to the pressure and temperature at the Earth's surface, all the ozone in the stratosphere would be contained in a layer around the globe just 1/3 cm thick).

Ozone in the stratosphere is important because it acts as a 'UV filter', cutting out shortwave ultraviolet radiation from the sun which otherwise would be damaging to plants and animals as they have evolved on Earth. Also, as a result of absorbing this short-wave highly energetic radiation, ozone causes a warming of the upper stratosphere, which influences global air motions.

The photochemical processes which form and destroy ozone in the stratosphere, and which thereby set up a dynamic equilibrium between destruction and creation, were first defined by Chapman over 50 years ago (Chapman, 1930). He showed that a series of reactions occurred, starting with the photodissociation of O_2 by sunlight at wavelengths below 246 nm, along the following reaction paths:

$$
\left.
\begin{aligned}
O_2 + h\nu &= O + O , \\
O + O + M &= O_2 + M , \\
O + O_2 + M &= O_3 + M , \\
O + O_3 &= O_2 + O_2, \\
O_3 + h\nu &= O_2 + O .
\end{aligned}
\right\}
\tag{3.1}
$$

The balance of these reactions, taking into account their reaction rates, the variation of solar radiation with height, and the mean atmospheric density (in the above equations 'M' represents an air molecule — i.e. predominantly N_2 or O_2), leads to a layer of ozone with a peak in the mid-stratosphere, as shown in Fig. 3.1.

In the 1970s, through pioneering work by Paul Crutzen, Harold Johnson, and others, it was realized that this set of equations was not complete, and could not actually reproduce the observed concentrations of ozone quantitatively: it was necessary to search for other ozone loss mechanisms. The important breakthrough

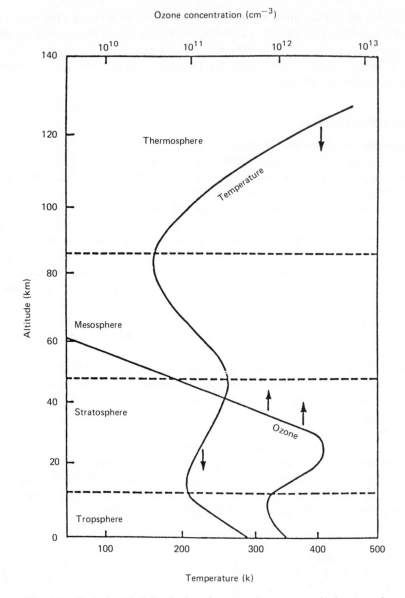

Fig. 3.1 — Typical vertical distribution of ozone and temperature in the atmosphere.

that then occurred was the realization that other chemicals in the atmosphere, albeit present in concentrations 1000 times lower than ozone itself, could be involved as catalysts in cycles which could make them highly efficient in destruction of ozone. For example, in the case of the nitrogen oxides, NO and NO_2, present in the atmosphere in the 10^{-9} to 10^{-8} range of relative concentrations,

$$NO + O_3 \quad = NO_2 + O_2 \, ,$$
$$NO_2 + O \quad = NO + O_2 \, , \qquad\qquad (3.2)$$
$$net \ O_3 + O \quad = O_2 + O_2 \, .$$

In other words, the net effect is that O_3 is reduced to O_2, and the NO and NO_2 are constantly interchanged, but not destroyed. Later work has revealed that chlorine, Cl, and chlorine monoxide, ClO, can play a similar role to that of NO and NO_2, as can OH and HO_2. Modern theories of ozone in the stratosphere have become very complex, with over 100 identifiable chemical reactions simultaneously in operation (see Brasseur and Solomon, 1984).

The growth in man-made pollution of the atmosphere by nitrogen oxides (e.g. from aircraft engine exhausts), and more seriously by chlorine-containing fluorocarbons (used in refrigeration, foam plastics and in aerosol spray cans) has given rise to fears that the delicate ozone balance might be shifted, with possibly serious consequences. Prior to the discovery of the Antarctic ozone hole, emphasis was placed by researchers on studying long-term ozone trends and links with such pollution. Then, reported in the scientific literature in 1985, came the sensational discovery of an apparently major decrease in ozone over a very short space of time (~5 years), which brought to our attention two things. First, how poorly we understood the atmosphere (none of our best theoretical models had any explanation for such a dramatic phenomenon); and, second, that environmental crises were not just a thing of the future, of long-term effects, but could occur on what is climatologically an instantaneous time scale.

3.1.2 The observations

The discovery of the Antarctic ozone hole is a delightful 'David and Goliath' story, which underlines the ease with which satellite observations of the Earth can be misinterpreted unless the utmost care is taken in their analysis.

In 1985, Dr Joe Farman and colleagues at the British Antarctic Survey announced in the journal *Nature* that their long-term series of measurements of total stratospheric ozone over the BAS base at Halley Bay in Antarctica showed that a huge loss of ozone was occurring each Austral spring (Farman *et al.* 1985). This loss had started in about 1979 and seemed to increase in magnitude each year up to 1984, when their publication was written. This annual loss occurred between September and November, was non-existent in data before about 1979 going back to 1957, and by 1984 was quite unambiguous compared with the errors of the measurements. Some of the data are shown in Fig. 3.2, which plots the total column of ozone (that is, the total amount in the stratosphere integrated over all heights) over the Halley Bay station (at a southern latitude of 76 degrees) each October from 1957 to 1984. Clearly, Farman and his colleagues, by a process of careful and patient measurement over many years, had stumbled on a phenomenon of major significance for the Earth's climate. Unless the changes could be accounted for by some so far unpredicted natural process (which of course might be no less serious for the climate), this seemed to be evidence of impact of man's activities on the natural environment which was irrefutable.

This, then, was the David of our story. The BAS team, in common with most

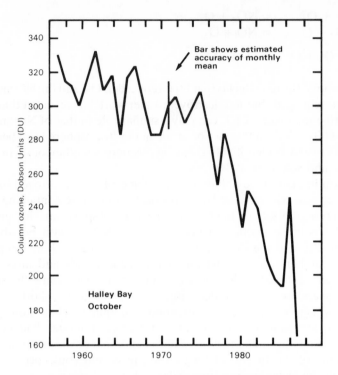

Fig. 3.2 — The October average overhead column of ozone observed from the ground at the British Antarctic Survey station at Halley Bay (76 degrees S).

atmospheric research groups in the UK, was small, and not particularily well funded. Nevertheless, the high intellectual level of British researchers has shown itself on many occasions, and this was one of them.

Goliath appears in the guise of the giant space agency, NASA. NASA operated a satellite instrument, the Total Ozone Mapping Spectrometer (TOMS), mounted on its polar-orbiting series of satellites, especially Nimbus. This had shown no such ozone hole developing over the Antarctic. What was the problem? Were the British data in error? Goliath quickly realized his mistake, and was in this case very open and generous in admitting that he had simply missed the phenomenon, even though it was in fact apparent in his data. Quite simply, ozone total column amounts below a certain value (about 250 Dobson units (1 Dobson unit = 1 milli-cm atm of column ozone)) had never before been observed anywhere around the globe, so the data processing programs in NASA's computers contained a threshold filter, which rejected data that fell below this level, since such low levels would be, on previous evidence, simply an error of measurement. Every time an ozone hole event occurred, therefore, the data processing system simply did what it was programmed to do, that is, it rejected the data as bad measurements. Fortunately the raw data had been preserved by NASA, and the agency quickly recovered by reprocessing without such a filter in operation, and were able to confirm the BAS results, and moreover were able to go on to produce valuable and informative maps of ozone distributions over the south polar regions like the one shown here (Fig. 3.3) and later in Chapter 5

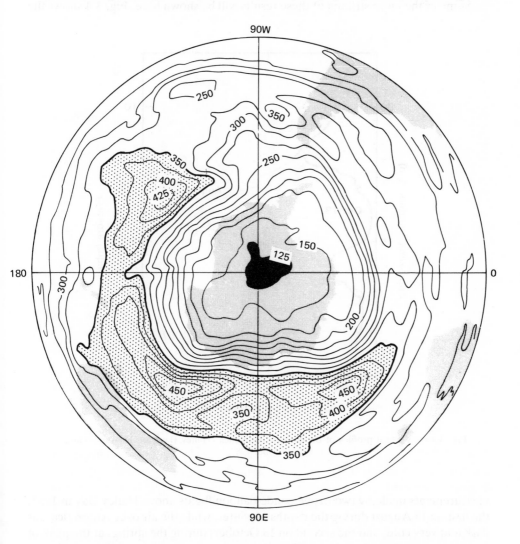

Fig. 3.3 — Observations from NASA's Total Ozone Mapping Spectrometer (TOMS) on Nimbus 7 for 5 October 1987, when ozone values over the whole Antarctic continent were below 200 Dobson units, and at the pole reached values below 125 D.U.

(Plates 7 and 8), which give a powerful synoptic view of the whole southern hemisphere every few days, and which taught us much about this phenomenon.

Other measurements have been made since this discovery, most notably in the context of some rapid-response campaigns to the Antarctic organized by NASA, NOAA, and other agencies. Here, the USA has regained its esteem by turning people and resources (including aircraft) in large quantities onto the problem, and thereby giving us a major new dataset by which to study the problem.

Some of the more striking of these results will be shown here. Fig. 3.4 shows the

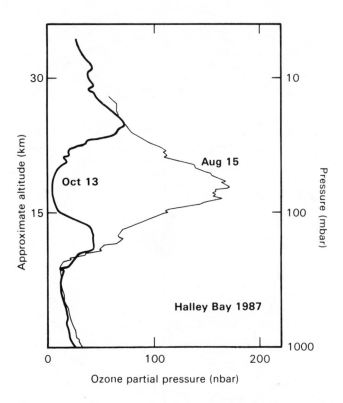

Fig. 3.4 — Vertical profiles of ozone over Antarctica (Halley Bay) for 15 August and 13 October 1987.

measurements made by two ozone-sonde balloon flights above Halley Bay in 1987, the first on 15 August during the depths of winter, while the air over Antarctica was dark and very cold, and the second on 13 October, during the spring, at the peak of the ozone depletion when almost 95% of the ozone between 14 and 23 km had been destroyed.

Airborne experiments give us more information on latitude variation, and on the trends in other atmospheric constituents such as water vapour, nitrogen oxides and chlorine oxides. Fig. 3.5 gives a schematic summary of the data observed. The 'normal' stratosphere seems to go through a transition region at about 65 deg. S, and then into a chemically perturbed region (CPR) on the poleward side of the transition zone. This CPR is characterized by (as well as low ozone in the spring), low H_2O, low nitrogen oxides, high values of chlorine oxides, and (not shown in the figure), ice clouds known as polar stratospheric clouds.

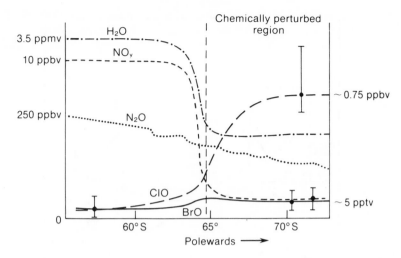

Fig. 3.5 — Illustrating the 'chemically perturbed region' and the 'normal' stratosphere outside it.

The extent of the CPR seems to be defined by the extent of the well-known, strong winter vortex which exists in the stratosphere during the months of darkness over the Antarctic (see Fig. 3.6). We shall go on in the next section to consider what these various observations are telling us about the processes responsible for the ozone hole. Later, in Chapter 5, we will show how the satellite observations, having missed the discovery of the effect initially, have fully redeemed themselves by producing wonderful global scale maps of ozone which help our interpretation enormously.

3.1.3 The theories

The presently accepted explanation of why the Antarctic ozone hole exists is not fully proven, and some worrying uncertainties exist, for example, a convincing explanation of why the ozone hole has grown so rapidly from year to year is still lacking. However, present wisdom (e.g. Solomon, 1988) relies on a series of logical steps which we will outline here. Of particular relevance to the argument are two things about the 'normal' stratosphere:

- Ice clouds are not thought to play a major role in the chemical balance: all chemical reactions of importance can be treated as being in the gas phase only, and not involving reactions on the surfaces of particles.
- To a significant extent, the ozone-destroying power of both the nitrogen oxides and chlorine neutralize each other by jointly forming rather stable 'reservoir' species, such as $ClONO_2$.

The series of processes currently thought to be responsible for the complete cyle of events giving rise to the ozone hole, at least under the conditions experienced in

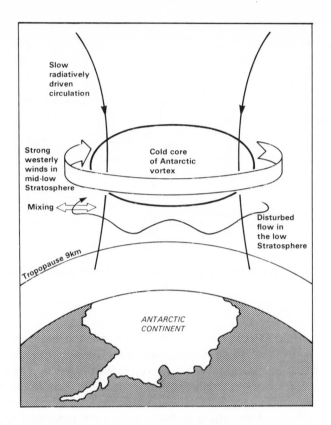

Fig. 3.6 — A schematic diagram of our present understanding of the Antarctic polar vortex in winter.

the Antarctic, are shown in Table 3.1. From this Table we can see that the overall effect of all the processes is to 'precondition' the Antarctic stratosphere so that when sunlight returns to the stratosphere after the long, dark polar winter, the net reaction

$$2 \times O_3 + h\nu = 3 \times O_2 \qquad (3.3)$$

occurs, i.e. the photolysis of O_3. Once the spring is well advanced, the high Sun induces a dynamical breakdown of the stable vortex, and the 'chemical containment vessel' as it is sometimes described is broken down, mixing of air with lower latitude air occurs, PSCs evaporate as the air warms, and the atmosphere returns to a more normal state.

Some important questions remain outstanding, however, some of which we can describe here:

- We have already noted that it is not at all clear why the appearance of the ozone hole over Antarctica has been so sudden and dramatic. The amount of available

Table 3.1 — Sequence of processes forming the ozone hole

	The series of processes currently seen as responsible are as follows:
Vortex formation	
	The formation of the winter polar vortex leads to isolation of cold, dark air over the pole for long periods.
	↓
Very low temperatures, PSCs	
	This leads to very low temperatures in the stratosphere, and the formation of significant amounts of polar stratosphere clouds (PSCs).
	↓
De-humidification, de-nitrification	
	This leads to low water vapour concentrations (de-humidification) and to removal of active nitrogen oxides by adsorption of HNO_3 onto the PSCs. These no longer neutralize the chlorine species.
	↓
Pre-conditioning	
	The PSCs also provide a reaction surface for transferring inactive chlorine to an active state ready to react once sunlight appears in the spring; an example reaction is:

$$HCl + ClONO_2 \rightarrow Cl_2 + HNO_3$$

(chlorine 'reservoirs')   Surface reaction   gas phase   left on the ice

	↓
Spring arrives: sunlight	
	The arrival of UV radiation photolyses the Cl_2 and leads to the formation of ClO:

$$Cl_2 + hv \rightarrow Cl + Cl$$
$$Cl + O_3 \rightarrow ClO + O_2$$

	↓
Ozone destruction by catalysis	
	What mechanisms could catalyse ozone destruction? Unlike at mid-latitudes, little atomic oxygen, O, exists at high latitudes. It is presently thought that the photolysis of the ClO dimer, Cl_2O_2, is the key reaction in a complex:

$$CO + ClO + M \rightarrow Cl_2O_2 + M$$
$$Cl_2O_2 + hv \rightarrow Cl + ClOO$$
$$ClOO + M \rightarrow Cl + O_2 + M$$
$$2 \times (Cl + O_3 \rightarrow ClO + O_2)$$
Net: $2O_3 + hv \rightarrow 3O_2$

Therefore, the overall effect of all these processes is to 'precondition' the Antarctic stratosphere so that when sunlight returns to the stratosphere in the spring, the net reaction

$$2O_3 + hv \rightarrow 3O_2$$

occurs, i.e. photolysis of O_3. Once the spring is well-on, the high Sun induces a dynamical breakdown of the stable vortex, and the 'chemical containment vessel' as it is sometimes called is broken down, mixing of air with lower latitudes occurs, PSCs evaporate, and the atmosphere returns to a more normal state.

chlorine in the atmosphere has, we know, doubled in the past 10 years or so, but unless some chemical threshold effect exists, beyond which ozone depletion becomes very rapid, it is hard to see that this doubling could cause such a rapid change. Changes in atmospheric temperature, humidity (owing to increasing methane levels, for example), or aerosol loading (owing to volcanic eruptions) have all been considered as ways in which the concentration and effect of PSCs could increase suddenly. It seems likely that a non-linear threshold effect in the chemistry is most likely to explain this effect, though no thoroughly convincing explanation of the sudden onset of the phenomenon has yet been achieved.

• An obvious question is whether the ozone hole can spread, in area or in depth. While the extreme conditions which occur over the Antarctic during winter and spring are not reproduced elsewhere on the globe, it is possible that more limited, but similar, effects may occur over the northern pole. Global effects of the Antarctic seasonal depletion are also not clear: the long-term effects of the annual loss of ozone over the Antarctic could include a spreading of the deficit by a dilution effect gradually to lower latitudes. Whether this happens obviously depends on the relative strengths of the various production and loss effects around the globe, in comparison to those locally over the polar regions.

• We have not been able to study aspects of stratospheric ozone at locations away from the poles in any detail in this introductory book, but we should note, in connection with the possible spread of the Antarctic ozone loss to lower latitudes, that a number of studies have been carried out to investigate whether there is evidence for a global depletion of ozone amounts. The best available reviews (see below for a short bibliography) seem to indicate a measurable decrease in the total ozone amounts between 30 degrees N and 64 degrees N for the winter months December to March, of between 2.3 and 6.2 per cent over the years 1969 to 1986. These results are based on measurements at a variety of locations around the globe, and are thought to be broadly consistent with what is predicted by computer models for the loss of ozone caused by the increasing levels of chlorine in the atmosphere, resulting from growth in the abundances of the chlorofluoro-carbons. To date, satellite measurements of global ozone do not extend over a sufficient time, nor do they have a sufficient accuracy, to allow trend analysis: this position should improve as satellite-based measurements improve. Two recent papers (Atkinson *et al.*, 1989: Proffitt *et al.*, 1989) add to the evidence for high and mid-latitude ozone loss outside the CPR.

• The question of how the polar ozone depletion might interact with other feedback processes and trends is also open. For example, could the anticipated (but not unambiguously detected yet) global greenhouse effect have any influence on the ozone hole problem? Or could a lowering of ozone concentrations lead to a cooling, which would strengthen the winter vortex and enhance the ozone depletion? Or could the 'quasi-biennial oscillation' (QBO) cause an alternation of the yearly 'bite' into the ozone amounts at the poles? Could the ozone hole phenomenon have occurred at earlier epochs? We have good evidence that the situation has been stable for at least some 20 years prior to the onset of the recent events, from the BAS data extending back to 1957: but we have no data whatsoever on stratospheric ozone earlier than this. (We do, however know from work by Stanford and others (e.g. Douglass and Stanford, 1982), that polar stratospheric clouds have been well known particularly over the southern polar ice cap for many years).

The theoretical study of the ozone balance over the polar regions, as well as more globally, is a rapidly developing subject, receiving widespread support especially in the USA, and engaging some of the best brains available in atmospheric science. It is therefore not possible in this introductory book even to attempt to be completely current on all aspects of this rapidly developing area of theory. Nevertheless, the reader should be aware of some of the major reports that have been produced, which will provide ample sources for further reading on the subject. For a very comprehensive review of 'pre-ozone hole' research on the stratosphere, the reader is referred to the documents published by NASA, WMO, and other agencies, under the title *The Stratosphere 1985*, available from NASA Headquarters in Washington, D.C. (NASA, 1985). More recently, since the realization of the Antarctic hole phenomenon, two important reports have been published internationally: first the Report of the Ozone Trends Panel, which, as the name suggests, considered the evidence for global changes in ozone amounts: *Present state of knowledge of the upper atmosphere 1988: an assessment report* (NASA, 1988c); and secondly the *Scientific Assessment of Stratospheric Ozone, 1989*, sponsored by the United Nations Environment Programme and the WMO (UNEP, 1989). These two documents are available from either NASA Head-quarters in Washington D.C., or WMO Headquarters in Geneva. In the UK, a government-sponsored team drawn from universities and institutes around the country and called the Stratospheric Ozone Review Group (1987; 1988) has published two reports so far (*Stratospheric Ozone* in 1987 and *Stratospheric Ozone 1988*), and is working on a third report to be published in 1990. These are all available from HMSO in London.

3.1.4 Relevance of observations from space

Quite clearly, quantitative global measurements of atmospheric composition on a long-term basis from space are an essential element in a global Earthwatch system. Equally clearly, the lesson that has to be learned is that in the interpretation of those measurements we must be careful not to eliminate classes of results, simply because they do not conform with our preconceived ideas of what should happen. Our Earth, its atmosphere and its climate are still capable of delivering some surprises to us, at fairly short order. Later on in this book, in Chapter 5, we shall show some examples of how new satellite data are assisting in the study of the ozone problem, both over the poles and more globally. Of particular relevance are measurements of ozone itself, and of the polar stratospheric clouds that we have seen play such an important role in the polar ozone issue. Space systems are also capable of measurements of many other chemical species in the stratosphere that are involved in the complex chemical cycles that maintain the balance of ozone.

3.2 THE GREENHOUSE EFFECT

In section 2.1.5 we discussed in a fairly elementary way the derivation of a radiative equilibrium temperature for a planet such as the Earth, on the basis of an optically grey planet with a certain albedo, or global reflectivity. We also saw that the existence of an atmosphere could lead to a surface temperature substantially higher than the global equlibrium temperature of the planet, owing to the blanketing effect of the atmosphere. The extreme example of this is Venus (see Table 2.2).

The greenhouse effect arises because the atmosphere is largely transparent to incoming solar radiation (ignoring the effect of clouds), while being quite heavily absorbing to outgoing thermal radiation from the planetary surface and the atmosphere. Detailed studies of the problem have been undertaken by several scientists, including Ramanathan *et al.* (1985), Kondratyev (1969), Manabe (1983), Mitchell *et al.* (1987), Hansen *et al.* (1988) and others. Both the pure radiation-transfer problem, assuming that the atmosphere can be represented by a static, non-convecting medium, and the coupled radiation-dynamical problem, in which the interactions of radiative heating effects and motions are considered, have been studied. The recognition of a climatic influence from changing atmospheric CO_2 amounts goes back, according to Manabe, to work by Arrhenius (1896) and Tyndall (1861) in the 19th century, and Callendar in 1938.

In this chapter we will examine this concept in somewhat more detail, look at some of the work that has been done to model the planetary greenhouse effect, and consider the role of space observation systems in monitoring and understanding the problem.

3.2.1 Radiative equilibrium in a real atmosphere

Fig. 3.7 shows a schematic representation of the atmosphere and the balance of radiative processes acting on it, in a very simple form. First we note that the total net incident radiation on the top of the atmosphere, which has a value of about 236 W m^{-2}, must be balanced by an equal outgoing radiance if the planet is in net thermal equilibrium, as the Earth must be when averaged over all the whole globe and over a sufficiently long period (e.g. one year). This may not be true instantaneously, for example there must be a net input of radiation locally during the day, and a net output during the night, but on average over a sufficient range of time and space this net balance must be true. Within the atmosphere, Fig. 3.7 tells us that some 76 W m^{-2} of energy is absorbed by the atmosphere from the incoming solar beam, and the remaining 160 W m^{-2} is absorbed by the surface of the planet, the land, oceans, and to a lesser extent (because of its high latitude) the polar regions. At the Earth's surface several processes occur, the most obvious being the re-radiation of energy owing to the temperature of the surface, in the infrared region. At a temperature of about 288 K (the equilibrium temperature we derived in Chapter 2 on a very simple basis was 280 K — see Table 2.2), this thermal radiation releases about 390 W m^{-2} into the atmosphere. A second process, the release of latent heat due to condensation and evaporation of water from the surface, is also important, and releases about

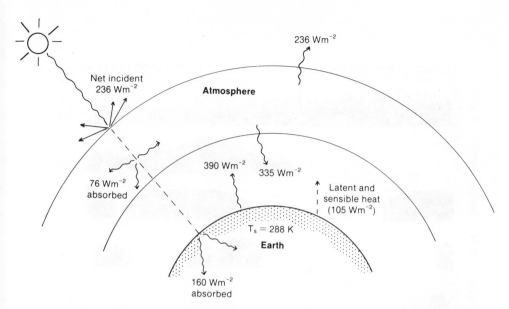

Fig. 3.7 — A simple model of the atmospheric radiation balance: a net incident shortwave flux from the Sun of 236 W m^{-2} is balanced by an outgoing equal flux in the infrared.

105 W m^{-2} into the atmosphere. Because it is spectrally absorbing in the infrared, the atmosphere emits about 335 W m^{-2} downwards from its lower, warmer layers, back towards the surface of the Earth. However, the 'topside' radiation to space is much lower, because it is being emitted generally from a much higher, colder layer of the atmosphere. This is an alternative way of looking at the 'greenhouse effect', of course : because of the spectral blanketing effect of the atmosphere, it is only the high, cold layers of the atmosphere which emit to space, and because of the low temperature the quantity of energy emitted is low. In the lower layers of the atmosphere, below the optically thick central portion, the net radiation is directed downwards, and because the temperature is higher, so is the emitted energy. Thus the effect of the greenhouse 'blanket' is to reduce the effective temperature of radiation from the atmosphere to space, that is, to reduce the energy loss from the planet, or in other words to warm the planet by limiting the escape of energy.

Of course, the greenhouse effect is a perfectly natural process which exists whether or not mankind is thought to be modifying the climate. It is the *enhancement* to the greenhouse effect resulting from man's activities that is liable to de-stabilize the natural balance. Many different gases in the atmosphere contribute to the greenhouse effect. All have in common the existence of spectral absorption bands in the infrared, sometimes quite weak ones. However, if these bands occur at wavelengths where the Planck radiation function peaks, or in spectral gaps of high transmission, their effects can be large. Fig. 3.8 shows the position of the absorption bands of a number of greenhouse 'active' gases in the mid-infrared spectral region, including in the fairly transparent 'window' region around 1000 cm^{-1}. At most other

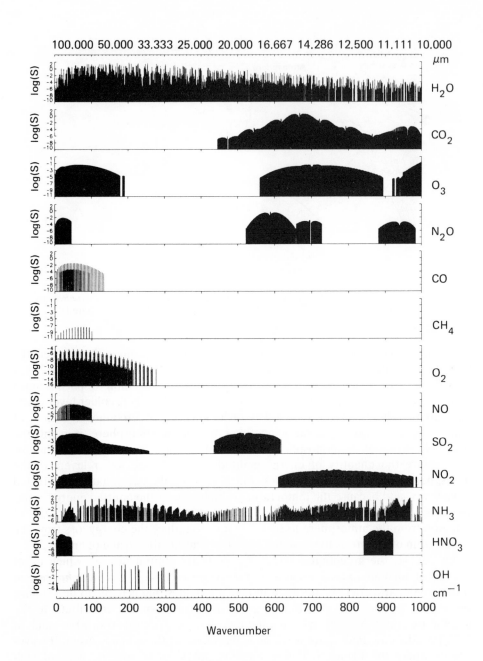

Fig. 3.8(a) — Positions and strengths (in arbitrary logarithmic units) of the absorption bands of numerous atmospheric gases, including 'greenhouse' gases, in the (a) 0 to 1000 cm^{-1} and (b) 1000 to 2000 cm^{-1} (microwave to infrared) spectral regions.

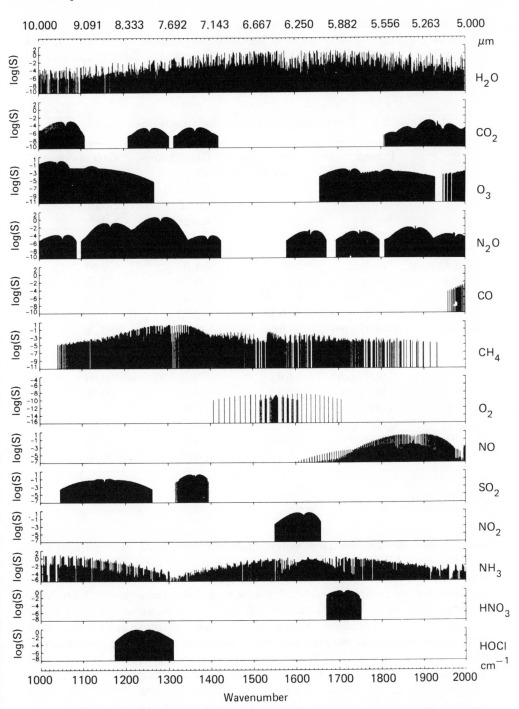

Fig. 3.8(b) — Positions and strengths (in arbitrary logarithmic units) of the absorption bands of numerous atmospheric gases, including 'greenhouse' gases, in the (a) 0 to 1000 cm⁻¹ and (b) 1000 to 2000 cm⁻¹ (microwave to infrared) spectral regions.

wavelengths in the infrared, the atmosphere is strongly absorbing, so that the occurrence of even weak absorption bands in the window region can have a substantial effect on the thermal balance. The relative effect of different gases on the greenhouse effect has been calculated by a number of authors. Table 3.2 gives some

Table 3.2 — Greenhouse effect of various atmospheric gases

Gas	Present concentration	Increase in concentration	Greenhouse effect for fixed cloud temperature (K)	Greenhouse effect for fixed cloud height (K)
1. H_2O	75%[a]	2-fold	1.03	0.65
2. CO_2	330 ppm[b]	1.25-fold	0.79	0.53
3. O_3	3.43 mm[c]	0.75-fold	−0.47	−0.34
4. $CFCl_3$	0.0001 ppm	20-fold	0.54	0.36
5. CF_2Cl_2	0.0001 ppm	20-fold	0.54	0.36
6. N_2O	0.28 ppm	2-fold	0.68	0.44
7. CH_4	1.6 ppm	2-fold	0.28	0.20
8. NH_3	0.006 ppm	2-fold	0.12	0.09
9. HNO_3	0.005 mm	2-fold	0.08	0.06
10. C_2H_4	0.0002 ppm	2-fold	0.01	0.01
11. SO_2	0.002 ppm	2-fold	0.03	0.02
12. CH_3Cl	0.0005 ppm	2-fold	0.02	0.02
13. CCl_4	0.0001 ppm	2-fold	0.02	0.01

[a]Relative humidity.
[b]Parts per million.
[c]Column density (mm at STP).

results from a one-dimensional model calculation by Wang *et al.*, and reported by the Soviet worker A. S. Monin (1986): two cases are quoted, one for a fixed cloud temperature in the model he has used, and the other for a fixed cloud height, in which the cloud-top temperature can vary. It can be seen that carbon dioxide, CO_2, has a major effect, in that a temperature rise of 0.53–0.79 K is predicted for just a 25 per cent increase in concentration. Other gases, while not as effective as CO_2 in creating greenhouse conditions, nevertheless cause significant effects: these include the fluorocarbons CF_2Cl_2 and $CFCl_3$, N_2O and CH_4 (methane). It will also be noted from the table that while most of the gases have a warming effect if they increase in concentration, ozone alone, O_3, causes an increased cooling. This arises simply because ozone is the only one of the gases which exists predominantly in the stratosphere. Because this is a region of cooling to space, an increase in ozone concentration would lead to an increased cooling to space, not a warming at lower levels as for the other gases.

Taking into account that a significant warming in the case of CO_2 is predicted for a rather small increase of some 25 per cent, we can understand the concern of climate researchers when we study how the concentrations of CO_2 in the atmosphere are changing. Fig 3.9 shows the results of almost twenty years of measurements in the

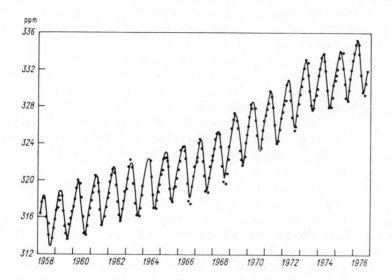

Fig. 3.9 — Measured changes in atmospheric CO_2 at Mauna Loa Observatory, Hawaii.

Pacific (Hawaii) of the concentration of atmospheric CO_2. These measurements have for a long time been considered to be representative of the background levels of CO_2 in the atmosphere, though recently this claim has been challenged. Nevertheless, the data indisputably show a steady year-by-year increase in CO_2, with a superposed annual cycle which is thought to be due to the annual cyle of photosynthetic activity in the northern hemisphere. With the sort of trends identified in this data already happening, it is perfectly possible to conceive of CO_2 increases of a large magnitude, even a doubling, occurring by the middle of the next century. We shall see below, when we look more closely at the results of model studies, that this can in principle lead to global warmings of up to 5 K, with extremely important consequences for the well-being of the planet.

Trends in other greenhouse gases, such as CH_4, N_2O, and the chlorofluorocarbons, also show increases with time as they are produced either directly (e.g. chlorofluorocarbons) or indirectly (e.g. methane) as a result of mankind's activities. Thus the greenhouse effect due to these gases is also thought to be on the increase.

The reader may have also heard of the threat from de-forestation. Apart from the very direct effects of soil erosion caused by removing trees in the world's great forests, these trees are also responsible for absorbing large quantities of CO_2 through photosynthesis: in a way, this process is a means of 'locking up' carbon in the form of wood and foliage. Without this absorber, the carbon, (in the form of CO_2) does not get taken up, and remains in the atmosphere. It is now believed also that not only forests but tropical and sub-tropical grassland is another major 'sink' for CO_2. The oceans, too, as we have seen in an earlier chapter, act as an absorber of atmospheric CO_2: we are not sure, however, of the true capacity of the global oceans to mop up CO_2, nor how near to saturation they may be, though both direct solubility of CO_2 in sea water, and take-up by sea animals such as plankton are thought to be important.

The risks of the greenhouse effect are therefore very real, and represent a huge threat to civilizations on this planet.

There is also a greenhouse effect due to clouds, as we have seen in Chapter 2. In the case of greenhouse gases, we can describe the phenomenon as a spectral filter acting on the radiation field, since it is the differential spectral absorption properties of the atmosphere that give rise to the phenomenon. In the case of clouds (see section 2.3), however, the process is one of spatial filtering, since it is the physical presence or not of a cloud that gives rise to the effect. The quantitative climatic effect of a cloud — i.e. whether à warming or a cooling occurs, and if so how large — depends strongly on the height of the cloud.

If we first of all consider a low, thick cumulus type of cloud, then this is highly reflective to incoming solar radiation, so that it quite efficiently prevents this radiation from heating the surface, i.e. a cooling effect occurs. On the other hand, the cloud is also an effective trap for infrared radiation leaving the surface, i.e. a warming effect is produced. The net effect is small, and since the cloud emits to space at a temperature close to that of the Earth's surface (because of its low altitude), the change to the column radiation balance due to the presence of the cloud is small, i.e. the net greenhouse effect of the cloud is small.

For high, thin clouds, however, the situation is quite different. Such clouds are poor reflectors of incoming sunlight, and so do not detract significantly from the incoming heating radiation. They are, however, quite good reflectors of upwelling infrared radiation from lower layers of the atmosphere. Because of their great height, the tops of the clouds are cold, and therefore radiate much less energy to space than lower, warmer clouds. Therefore, these high clouds, contrary to imme- diate intuition, are in fact highly effective greenhouse blankets because they allow in much of the incoming radiation, yet radiate to space at quite low effective tempera- tures, i.e. they act to minimize energy loss to space.

3.2.2   Observed variations in surface temperature

What of the reality of global warming? A number of analyses of the trends or variations in surface temperature have been carried out over the years, but the results have often been ambiguous, and have not always given a clear conclusion. Nevertheless, three distinct classes of observational data can be identified, which can be discussed separately.

First, we have prehistorical data, based on specific scientific analysis techniques such as carbon dating and oxygen isotope analysis, or on the analysis of ice or sea- floor cores for pollen or insect densities. Second, we can utilize indirect measures which have been made during recorded history but before the advent of the use of scientific instruments and measurements made under controlled conditions, such as written or painted historical accounts, records of crop yields, or nautical records. Thirdly, in the period from about 1700 AD onwards, data from modern instruments have been available. Fig. 3.10 shows examples of estimated climate trends obtained from all three classes of source, from which it must be concluded that there have been

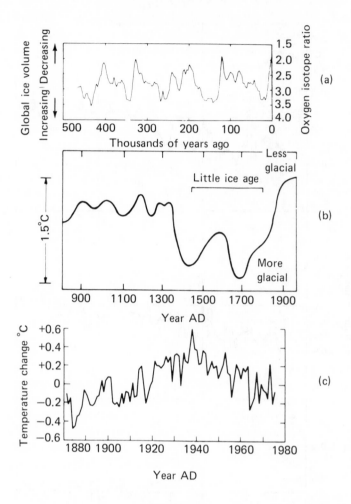

Fig. 3.10 — Climate over the past 500 000 years estimated from isotope and other measurements.

significant variations in surface conditions of the climate system over a considerable period of time. It is believed that there are a number of causes of such variations, such as changes in the Earth's orbit around the Sun, which might explain the larger, longer-term effects, but, solid theories to account for most of the variations are sparse.

The enhanced greenhouse effect, of course, affects only the period since the industrial revolution, since only since then has the output of CO_2 and other gases from fossil fuel burning risen dramatically. Fig. 3.11 shows a more recent analysis by scientists at the University of East Anglia in the UK, which, after a number of corrections to the data (see Jones, P. D. *et al.*, 1986) indicates a fairly clear trend of increasing temperature with time during the past 100 years over land and ocean regions, in both the northern and southern hemispheres (curves a and c), and, of

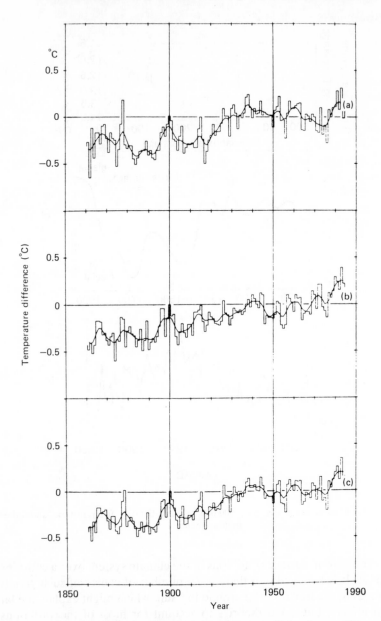

Fig. 3.11 — Measured variations in surface air temperatures during the past 130 years, in the northern (a) and southern (c) hemispheres, and as a global average (b), taken from land and oceanic measurements.

course, globally (curve b). Not all analyses show the same unambiguity in their results, and it is important to obtain a consensus between scientists to establish the best estimates of trends. Of course, the existence of a demonstrated trend still does not prove connection with the greenhouse effect, and might be due to other, possibly natural, variations. Nevertheless it is imperative to establish the true nature of the observed trends.

Recent work by Kuo *et al.* (1990) has shown a very strong correlation between atmospheric CO_2 and global temperature, using data for CO_2 between 1958 and 1988, and surface temperature data from 1880 onwards. Using sophisticated statistical analysis methods (multiple-window time series analysis) which are appropriate to short period data sets for estimation of trends and power spectra, these workers found a significant correlation between the two data sets over the last thirty years. This might be thought to be the first unambiguous detection of the enhanced greenhouse effect, except that the authors report that changes in CO_2 content lag those in temperature by about 5 months. A simple enhanced greenhouse effect would be expected to lead to the reverse. This result is still not fully understood, and represents the state of the art in trying to detect CO_2-induced enhanced global warming, at the time of writing.

3.2.3   Results of model simulations

We shall now take a look in more detail at what the results of computer model studies tell us about the greenhouse effect. Of course, we should be careful to realize that even the most sophisticated model can only be an approximate representation of the truth, and can indeed give misleading results: nevertheless, we can learn a great deal by investigating processes and how they interact with one another using computer models. In what follows, we shall call on the work particularly of Ramanathan and co-workers on one-dimensional studies, and of Manabe and co-workers in three-dimensional coupled atmosphere–ocean studies.

3.2.3.1   *One-dimensional models*

Ramanathan (1976) has used what is known as a one-dimensional radiative–convective model as a basis for much of his research. This means that the model considers one vertical dimension only, viewing the variation of parameters such as radiative energy flux and trace gas composition as a function of height. Variations with latitude or longitude are not treated by the model, but are compounded into one global average of any parameter for each altitude. Dynamical effects are treated as one-dimensional approximations of the real three-dimensional motions of the atmosphere by 'parameterization', that is, by an approximation equation which is adjusted to simulate the real atmosphere's dynamics as closely as possible. This sort of model is, of course, time dependent, and is particularly useful for studies of processes which do not depend to first order on geographical location, and for investigations of time variability. In the one dimension of the model the effects of radiative exchange are computed using spectrally averaging 'band models' of the atmospheric absorption spectrum, and vertical convection is parameterized using coefficients to represent the strength and direction of convective exchanges. This particular model includes a fairly representative approximation for the exchange of heat and moisture between the surface and the lower atmosphere, not always included in 1-D models. Chemical processes in the stratosphere are included, but not climate chemistry effects in the troposphere; stratosphere–troposphere temperature/water vapour feedbacks are also not included.

The rate of growth of various greenhouse gases in the atmosphere was thoroughly investigated by Ramanathan, and Table 3.3 shows some of the growth rates assumed in his study. It can be seen that increases of 30 per cent in CO_2, 25 per cent in N_2O, and up to a factor of 10 in chlorofluorocarbons are thought possible by the year 2030, only 40 years away from us now.

Using these growth rates, and repeating runs of his computer model for the two extreme values of concentration (1980 and 2030), Ramanathan obtains an estimate for the growth in the greenhouse warming due to these increases in composition. The results are displayed in Fig. 3.12, which shows the cumulative growth of the greenhouse effect as each gas is added. While the dominant effect is indeed due to CO_2 (as was recognized by Tyndall and others in the 19th century), we can see that the combined effect of all the other greenhouse gases is in fact about the same magnitude as that due to CO_2 alone. Overall, the model predicts an increase in global surface temperature of about 1.5–2.0 K, which would be a serious effect in terms of weather patterns and climate. Perhaps more serious would be the fact that, even assuming the most extreme preventative measures now, the years after 2030 would probably see a further growth in warming. While we must always remember to treat the results of model calculations as only an approximation to the truth, it is beyond question that by now the number of different studies that have confirmed the likelihood of an enhanced global warming due to this effect is so large that we must treat this threat to the world's climate very seriously indeed.

It is also interesting to note the predictions of Ramanathan's model when it comes to an analysis of how the greenhouse effect varies with height in the atmosphere. While we have already seen that the model does not include all the important feedback processes properly, it was found that the effect of adding CO_2 and the other gases, while leading to a warming in the troposphere, leads to a cooling in the upper layers of the atmosphere. This is caused quite simply because at these levels, with little absorption of radiation at higher altitudes, these gases radiate to space, as part of the global energy balance that we discussed earlier. This property of a greenhouse cooling at high altitudes in the atmosphere is shown in Fig. 3.13, where the surface warming changes to a cooling above the tropopause, becoming quite a substantial effect (a cooling of several degrees) in the mid-stratosphere above about 30 km.

3.2.3.2 *Three-dimensional models*
A number of groups worldwide have attempted to put together far more complex and sophisticated computer-based models than those used in the studies described in the preceding section. Though these models represent an attempt to give a considerably more detailed description of the climate system, they are also much more unwieldy than the lower dimensional models, and are only available to a few very well-endowed groups around the world. Rather than being the panacea for all modelling problems, we should rather regard these super-models as a part of our battery of tools to study the climate, having particular strengths and weaknesses. Thus they are essential when we need to simulate the complex interaction of many different processes as accurately as possible; however, they are so expensive and time-consuming to run that they are of far less value than the one-dimensional models discussed above for studies of individual processes (see section 2.7).

As an example of three-dimensional global climate models, we will concentrate our attention on the work of Syukoro Manabe and his colleagues (see Manabe,

Table 3.3 — Growth in some greenhouse gases, 1980–2030

Chemical group	Chemical formula	Average residence time (years)	1980 global average mixing ratio (ppb)	2030 estimated average mixing ratio (ppb)	Comments
Carbon dioxide	CO_2	2	339 000	450 000	Assumes 2.4% p.a. increase for the next 50 years
Nitrogen compounds	N_2O	120	300	375	Combustion and fertlizer sources
	$(NO+NO_2)$	0.001	0.05	0.05	Concentration variable and poorly characterized
Chlorofluoro-carbons	CF_2Cl_2	110	0.28	1.8	Depends strongly on degree of regulation
	$CFCl_3$	65	0.18	1.1	
	$CHClF_2$	20	0.06	0.9	
	CCl_4	25–50	0.13	0.3	
Hydrocarbons and others	CH_4	5–10	1650	2340	Based on recent measurements of trends
	CO	0.3	90	115	No trends identified to date
	H_2	2	560	760	
Ozone (troposphere)	O_3	0.1–0.3	25–70	30–100	Small trend has probably been measured

1983). This group has over the years refined the design, first, of atmospheric general circulation models, and then coupled atmosphere–ocean models. Fig. 3.14 shows the general structure of one particular version of a model, used to study the steady-state climate system, and the effect of the ocean on the climate. The model comprises three basic subsystems: (i) a general circulation model of the atmosphere, (ii) a continental surface heat and water balance model, and (iii) an ocean static, mixed layer model. The atmospheric model provides the basis of the whole scheme, and provides estimates of surface pressure, temperature, moisture and other parameters based on the fundamental equations of motion, thermodynamics and continuity. The amounts of solar radiation, ozone and CO_2 are precribed, i.e. the calculation is not fully interactive of these parameters. Condensation is assumed to occur, with consequent rain or snow fall at prescribed temperatures. Reasonably realistic allowance for the interaction of this atmosphere with the underlying land surface, and with the ocean is made in the model.

Figs 3.15 and 3.16 show results of the study carried out by Manabe and his colleague Stouffer using this model, in the late 1970s. The first of these shows a latitude–height cross-section of the zonal mean difference in annual mean tempera-ture obtained by the model when it is run including normal CO_2 amounts ($1 \times CO_2$) and four times this level of CO_2, with all other specified parameters remaining the same ($4 \times CO_2$). As predicted by the simpler models which we considered earlier, a warming of the troposphere is predicted, with a corresponding cooling at higher altitudes in the stratosphere. Some interesting further details are apparent, however, which could not be predicted by the simpler models. For example, the surface warming is seen to be considerably enhanced at high latitudes, because of positive

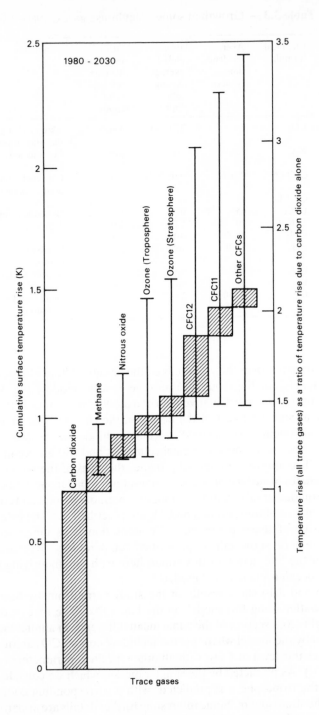

Fig. 3.12 — Predictions of cumulative global warming in the period 1980–2030 due to a number of 'greenhouse' gases, from a one-dimensional radiative–convective model.

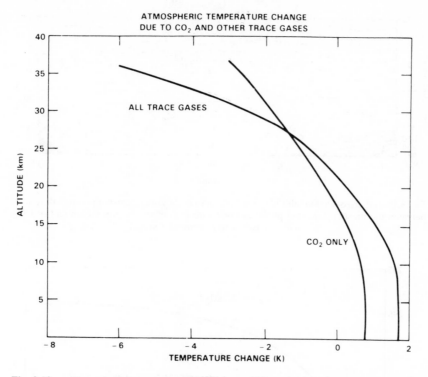

Fig. 3.13 — Atmospheric temperature change due to CO_2 and other 'greenhouse' gases, as a function of altitude.

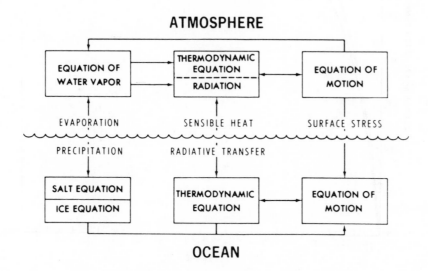

Fig. 3.14 — Basic structure of a coupled ocean–atmosphere model.

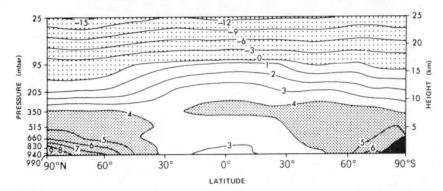

Fig. 3.15 — A cross-section (altitude versus latitude) of the zonal mean temperature difference between the present-day atmosphere and one with four times the present amount of CO_2, calculated using a large computer model.

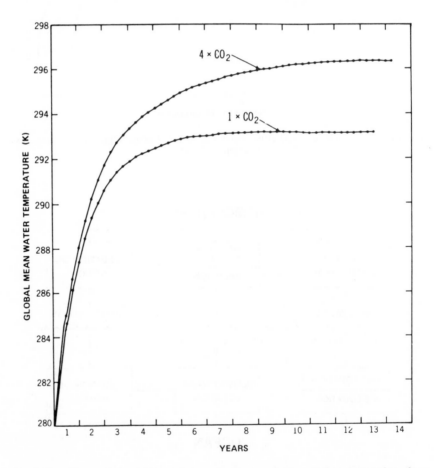

Fig. 3.16 — Time variation of the global mean water temperature at the ocean surface, from Manabe's $4 \times CO_2$ experiment.

feedback effects due to the highly reflective ice and snow cover (thus, an increase in temperature leads to an increase of melting, and a decrease of snow and ice cover: this in turn leads to an increase in the absorption of incoming solar radiation by the now darker surface, and thus an enhanced warming. It is suspected, however, that this positive feedback may be to some extent compensated by increased cloud cover resulting from increased available moisture, but this is not accurately modelled in this particular model).

At lower latitudes, the effect of the CO_2 warming is distributed throughout the troposphere by moist convective processes, which probably also account for the slight increase of the warming with height seen in equatorial regions. We should note that similar increases of the temperature effect with height do not occur at high latitudes simply because there the existence of stable stratification in the atmosphere inhibits vertical convective redistributions of the energy available from the greenhouse warming.

One further interesting effect to notice from Fig. 3.15 is that the polar warming enhancement in the southern hemisphere is smaller than that in the north. A moment's thought will make the reason clear: since the positive feedback in the north has to do with the creation of a larger area of low albedo due to the melting of the rather thin layer of sea ice, it is clear that in the south, with much thicker continental ice sheets than in the north, this positive feedback is considerably weaker.

The global average surface warming that Manabe and Stouffer obtained from this work was +4.1°C, for a fourfold increase in CO_2, implying that for a doubling of CO_2 (which has become a popular 'standard' case by which to intercompare different results) the rise would be about 2°C. In order to give the reader some idea of the consensus of results which are now becoming available, we show Table 3.4, which is

Table 3.4 — Global CO_2 warming: several results for a doubling of CO_2

Type of model	Reference	Geography	Sea-surface temperature	Insolation	Cloud	ΔT (K) doubling
Interactive ocean						
	Manabe & Wetherald (1980)	Idealized	Predicted	Annual	Predicted	3.0
	Wetherald & Manabe (1981)	Idealized	Predicted	Annual	Prescribed	3.0
	Manabe & Stouffer (1979, 1980)	Realistic	Predicted	Seasonal	Prescribed	2.0
	Hansen et al. (1981)	Realistic	Predicted	Seasonal	Predicted	3.5
Non-interactive ocean						
	Gates et al. (1981)	Realistic	Prescribed	Seasonal	Predicted	0.3
	Mitchell (1979)	Realistic	Prescribed	Seasonal	Prescribed	0.2

taken from a review by Manabe (1983), and which reports results from a number of groups on the effect of doubling CO_2. Broadly we can see two classes of findings, those with interaction between the atmosphere and the ocean (which we should, of course, expect to be the more realistic), and those without this interaction. The results of the models including an interactive ocean are all very close to one another, and imply some degree of 'rightness' as a result. However, we should note that to some extent these different models are not entirely independent. The different groups are working on the basis of a common pool of knowledge which exists in the scientific literature, and usually cooperate closely, so that some new development or idea in one group very soon becomes known to all their peer-group researchers. It is entirely possible that some new feedback process may at any time be discovered, which might radically change the results of this type of calculation — in which case we might expect this change to propagate to all other research groups, through the process of normal scientific communication, quite quickly. This view is not intended to undermine the enormous power and value of these model calculations, but only to try to give the reader a balanced perspective on how to judge their results.

Other interesting effects which are predicted by the large 'general circulation models' (GCMs), as they are known have also been found. It is observed, for example, that over Arctic regions the CO_2-induced warming shows a strong seasonal dependence, being at its strongest in the winter. This contrasts with the situation at low latitudes where very little seasonal dependence is predicted. Also, the quadrupling of CO_2 is found, not surprisingly, to increase rates of both evaporation and precipitation, as one might expect in a generally warmer environment. In the cryosphere, the amount of sea ice decreases, and the snowmelt season arrives earlier. At non-polar latitudes, water runoff rates increase, but most dramatically as far as human well-being is concerned, there are shifts in belts of moist and dry soil in the northern hemisphere. The consequences of such predictions for agriculture and potential crisis management are evident. Recent work by the Manabe group (Stouffer *et al.*, 1989) points to a slower warm-up in the SH than the NH, owing to the greater influence of the oceans in the south; and work by the UK Meteorological Office (Mitchell *et al.*, 1989) has indicated that a substantial negative feedback effect (i.e. a reduction of the greenhouse warming) may arise because models have so far inadequately described the effect of changes of state of water as clouds are formed and dispersed. It is clear that we are not yet in a position where we can predict global warming effects with any real accuracy.

3.2.4   Relevance of observations from space

It is still unclear what will be the most sensitive indicator of the onset of the greenhouse effect. It may be that direct observations of global mean surface atmospheric temperatures such as those shown in Fig. 3.12 may be the most direct and unambiguous indicator, or some other parameter such as stratospheric temperature decrease as shown in Fig. 3.13. Scientists are not yet agreed on such matters, facing as they do the task of trying to understand a system of great complexity. What must be indisputable is that we must strive to make better and better observations of the state of our climate, and develop the underlying understanding to be able to interpret those observations precisely. Satellite measurements have an extremely important role to play, owing to their global coverage, in making high-accuracy observations of parameters like ocean surface temperature, global cloud coverage, stratospheric temperatures, polar ice distributions and extent, and land surface vegetation and soil moisture. The difficulties of making many of the measurements from space are very great, but the importance of obtaining accurate global fields (i.e. spatial distributions) of these and other parameters is vital, and so the challenge to us is to overcome these difficulties.

3.3 'EL NINO'

The El Nino phenomenon is not a process with a clear beginning and end, at least not on the basis of our present understanding. It is a system of several coupled phenomona, each interacting with and driving the others, and what is not clear is just how a particular cycle of events is triggered: which starts up first, the chicken or the egg? Vallis (1988) has recently attempted a definition of an El Nino event, declaring at the outset that we must allow for the possibility that the 'events' are simply manifestations of a sustained oscillation process, albeit of some complexity and irregularity. Vallis states:

(1) El Nino is the occurrence of an anomalously warm pool of water in the eastern equatorial Pacific Ocean. The event lasts for a few months.

(2) Concurrent with the ocean warming, an atmospheric event occurs, namely a notable weakening of the trade winds. As one indicator, the sea level pressure difference between Darwin and Tahiti is correlated with the oceanic events.

(3) The events occur aperiodically, with intervals of between 2 and 11 years, but typically they are 2–5 years apart. There have been nine or so events since 1945, when reliable records began, with large events in 1957, 1965, 1972, and 1982. There is evidence for El Nino events for over 400 years (Quinn et al., 1987).

(4) The event is phased-locked to the seasonal cycle, normally reaching its maximum amplitude around Christmas time. However, large variations can occur, notably in 1982–1983.

(5) Similar events, if they occur at all, are much weaker in the Atlantic and Indian Oceans.

(We could also quote Hastenrath (1988) who records 76 events since 1726, based on work by Rasmussen.)

So, now we know! Or do we? The reader can probably deduce from these statements just how difficult it is in the case of the El Nino to give a simple 'pat' description of what is going on, as we can for many other phenomenon such as the greenhouse effect or the ozone hole. The El Nino phenomenon is much more entangled, more complex, and more difficult to provide with a simple explanation. Nevertheless, Vallis's definitions are very useful in focusing our thinking on cause and effect, and we can expand on this basis in looking at the phenomenon in more detail.

3.3.1 The phenomenon

The term 'El Nino' is Spanish for 'the little boy', meaning the Christ-child, a reference to the occurrence of the event, when it happens, being around Christmas time. The effect on the local fisherman is far from a happy one, however, and quite out of keeping with the Christian message at this time of year. For the fishermen, catches off the coast of Chile are severely depleted, because of the reduction of upwelling of deep cold water off the coast, which normally carries important nutrients and life-forms which provide food for the fish. Indeed, Hastenrath (1988) describes the local environmental/food production problem which follows from an El Nino as an ecological 'catastrophe'.

The behaviour of the atmosphere is intimately linked with that of the ocean in the El Nino. Under 'normal', i.e. non-El Nino, conditions, the warm waters of the western Pacific and the cool waters of the eastern Pacific are associated with strong surface easterlies, rising air in the west, and a strong upper-air return air flow, descending again in the east. This circulation cell has been well-known for some time and is known as the Walker circulation after Sir Gilbert Walker, who in the 1920s and 1930s developed the concept of the 'Southern Oscillation', which refers to a remarkable anti-correlation between the surface pressure in the south-east Pacific with that in the region of Indonesia: we will say more about this later, but for now, the circulation scheme which we have just outlined exists during one phase of the SO phenomenon. The prevailing easterly surface wind tends to build up a 'head' of water in the western Pacific, while in the atmosphere, vertical convection in the west is strong, and subsidence over the eastern Pacific is marked. Fig. 3.17 shows a cross section taken along the equator for two months, January and July, averaged for the years 1949–1953, and indicates the Walker circulation cell which occurs during normal non-El Nino years (taken from Bjerknes, 1969).

The pattern of sea-surface temperature during normal years is shown in Plate 2 in the upper frame, quite clearly demonstrating the west-to-east temperature gradient which normally exists, and which is associated with cold, upwelling water off the coast of South America, bringing nutrients to the surface waters from the ocean deeps. The lower frame of the diagram shows the difference between an El Nino year and a normal year: a warming of the eastern Pacific occurs, and the interconnected changes to atmospheric and oceanic circulations that have just been described take place. Though the temperature changes, of order 1 K or less, may seem to be small, they represent huge changes in energy balance across the vast extent of the Pacific since they occur as correlated, uniform changes over very large areas of the ocean surface. The data in Plate 2 are taken from satellite observations; we will consider the utilization of such spaceborne observations again in Chapter 5.

The El Nino event occurs when the west–east gradient of surface water temperature weakens. Then the easterlies slacken and the head of water in the west is released, and water 'sloshes' towards the east (speaking somewhat loosely, of course, of a phenomenon which occurs across the full breadth of the Pacific Ocean — some 3000 km and more). The pressure gradient between the eastern south Pacific and Indonesia is weakened. The weakened surface circulation causes a large reduction in the rise of cold deep water in the east. The release of the head of water in the west may be accompanied by the induction of equatorial Kelvin waves which propagate to the eastern extremity of the Pacific, taking about 2–3 months to

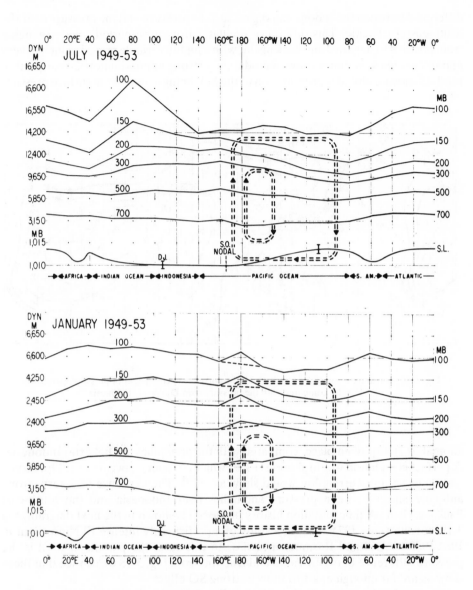

Fig. 3.17 — Illustrating the Walker circulation over the equatorial Pacific Ocean.

arrive there from the west, contributing to the warming of the surface waters. The change of temperature plus the high rainfall also associated with the event means ecological disaster.

3.3.2   Related phenomena
3.3.2.1   ENSO
We mentioned above that the El Nino oceanic effect is related to the Southern Oscillation. The SO is a major variation in pressure patterns, measured as a

difference between the Indonesian region surface pressure and the pressure over the south-eastern Pacific. This difference oscillates in a quasi-periodic way, swinging from positive to negative and back again, with a highly variable period, which in very marked events sometimes shows a moderately strong approximately 2-year period. Fig 3.18 shows a global representation of the SO, being a map of the surface pressure

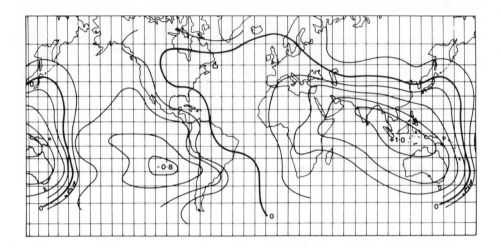

Fig. 3.18 — Walker's 'Southern Oscillation': the correlation coefficient of pressure variations worldwide, compared with the pressure over Djakarta, Indonesia.

anomalies at any given place, compared with the anomalies at Djakarta, Indonesia. A perfect correlation would show as +1.0 (which, of course, occurs over Djakarta), and an exact anti-correlation would show as −1.0. The existence of a very strong anti-correlation between pressure anomalies over Indonesia and the south-east Pacific is clear. A time variation history of the SO is shown in the next diagram, Fig 3.19, for the years 1974/1983, in a study of a particularily intense El Nino event in 1982/1983 by Gill and Rasmussen (1983). In this case, the parameter used is the pressure difference between Tahiti and Darwin: from Fig. 3.18 we can see that these stations are far enough apart to show a strong SO effect.

The SO has been found to correlate well in time with the El Nino, giving rise to the recognition of a coupled phenomenon, the 'El Nino–Southern Oscillation', or ENSO. The study by Gill and Rasmussen already quoted showed that the large and sudden drop in the Southern Oscillation Index (SOI) in late 1982 was accompanied by a very large warming event in the surface waters of the south-eastern Pacific, as shown in Fig. 3.20. The 1982/1983 event was a very clear example of ENSO activity. The rapid rise in temperature shown in Fig. 3.20 marked the onset of the El Nino proper, and in that case arrived some 3 months after the mid-1982 SOI fall, and the consequent collapse of the normal equatorial easterly winds. Following the onset of

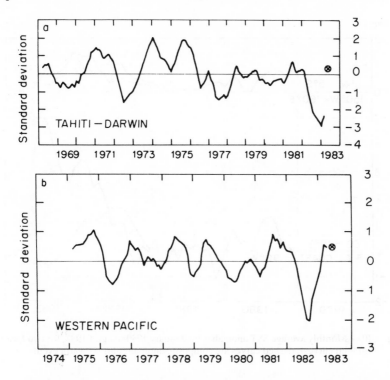

Fig. 3.19 — The time-history of (a) the Tahiti–Darwin pressure anomaly (a measure of the Southern Oscillation Index, SOI), and (b) the average easterly wind speed anomaly over the western equatorial Pacific, illustrating changes during the 1982/1983 El Nino.

the El Nino, there were record floods and rainfalls in Ecuador and Peru, the visit of HM the Queen to California was a wash-out, droughts occurred in Australia, and a major ecological crisis was occurring in many parts of the world (see Bigg, 1990).

3.3.2.2   Weather disturbances

Continuing the story from the previous section, the enhanced rainfalls over the eastern borders of the equatorial Pacific, which were so much in evidence in 1983, are accompanied by increased dryness in the western Pacific, simply because the changed circulation, which gives increased uplift of air in the east, and consequent rainfall, also gives a weakened uplift in the west, i.e. increased subsidence of air and dryness. Thus drought conditions tend to prevail during an ENSO event in the greater Australasia region.

Other connections have been deduced. The Indian monsoon tends to produce better rainfall to that thirsty continent during the 'normal' phases, and to be seriously disturbed during an ENSO event. Upper air patterns are disrupted, with the upper troposphere warmer during an event than normal, and with the upper westerly equatorial return flow of the Walker circulation weakened. Such episodes can in principle lead to significant variations in the supply of water vapour to the strato-

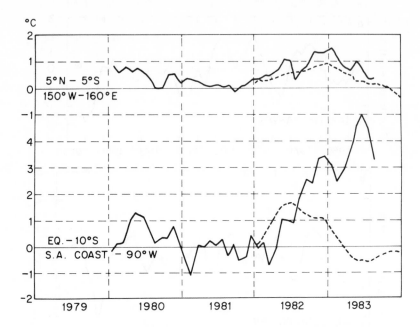

Fig. 3.20 — Monthly average SST anomalies for January 1980–August 1983, for two locations
(solid curves); dashed curves are composite anomalies for six warm events.

sphere, as I have noted elsewhere (Harries *et al.*, 1983). This leads to the very
interesting conclusion that ENSO-type events could drive the variability in stratospheric humidity, and also in various chemical processes which are affected by the
supply of chemicals from the troposphere to the stratosphere (see section 3.1). Thus,
one speculates, a significant link could occur between ENSO and stratospheric
chemistry/ozone balance, at least over equatorial regions.

3.3.2.3   Teleconnections

Such considerations have led to the concept of 'teleconnections', that is, connections
between different processes, often remote from one another, on a global basis.
Bjerknes, in papers published in the 1960s, discussed this idea, and gave some early
descriptions of many of the connected phenomena of pressure changes, air flow,
precipitation, rainfall, drought, and general environmental mayhem that we have
actually observed during the past decade. In this phenomenon, in contrast to the
problems of stratospheric ozone and greenhouse warming that we have discussed
earlier in this chapter, these 'teleconnected' ENSO-type events seem to be a natural
phenomenon which we are struggling to understand. So far, it seems that no-one has
suggested that such events are directly influenced by man's activities, unless a
gradual raising of global mean temperatures, or the consequent slight shifts in
weather patterns, are thought to exert some direct influence on the frequency or the
onset of ENSO events.

3.3.2.4   Periodicity

The question of periodicity of El Nino episodes obviously arises. Are there any regularities in these occurrences, however well disguised they may be? Do the processes involved permit any predictability? Can we develop our ideas enough that we could give warning to people sufficiently in advance for some of the terrible consequences of these events to be avoided by the populations of the countries most afflicted?

Gill and Rasmussen (1983) state that ENSO events occur at intervals of 2–10 years. Vallis (1988) reports 2–11 years, but says that typically the events occur from 2–5 years apart. Vallis also notes that there have been nine or so events since reliable records were started in 1945, with particularily large events in 1957, 1965, 1972 and 1982. Gill and Rasmussen have studied the 1982 event in some detail, and the 1969 paper by Bjerknes takes the 1957 event as an example. Other authors have analysed other episodes. Vallis also quotes the work of Quinn et al. (1987), which gives evidence for events over a 400-year span, and which shows that over this period El Nino events have tended to occur at intervals of about 3–4 years or 7–8 years.

In a recent paper, Hanson and co-workers (1989) have extended the analysis of the Quinn data. They take this data set over the period 1525–1987 (463 years) and identify 45 years in which strong El Nino events occurred. They have then subjected this time series to a statistical analysis which allows the identification of two major frequencies in this nonlinear, periodically varying phenomenon. They find two dominant periods emerge from the data, of 6.75 years and 14.0 years. They note in passing that they also see evidence for the harmonics of the 6.75 year period, and that the second harmonic of this period is equal to the period (2.2 years) of the quasi-biennial oscillation, the well-known reversal of stratospheric winds. Hanson et al. also notice from their data that the two periods tend to 'clump' together into 95-year blocks, with 14-year strong El Nino occurrences dominating every other 95-year block, interspersed with 95-year blocks in which the shorter period dominates. They speculate that this 'mode-switching' may be a frequency modulation (or beating) of the two periods.

Hanson and colleagues conclude their work with a careful disclaimer, that the statistical analysis which they carried out does not, of course, provide insight into the physical mechanisms which determine the onset of events. Indeed, they state that the fundamental cause of the quasi-periodic nature of these events remains unknown. This brings us on to a final comment on mechanisms of triggering ENSO events.

3.3.3   Final comments on El Nino

To conclude this section on the El Nino, we return to the work of Vallis (1988). In this paper, he shows that the basic phenomenology of the El Nino can be fairly well simulated by rather a simple model which includes only a minimally accurate representation of the interaction between the ocean and the atmosphere. In fact the model he discusses includes a simple feedback between near-surface ocean temperatures and the surface zonal wind, which is then allowed to react back on the ocean surface to drive advection of the surface waters, and thence the temperature field. This basic model produces two alternative stable states, one with higher temperatures to the west and a westward current, the other being the converse of this. If finite amplitude perturbations are introduced through stochastic (i.e. random) forcing, then switches between these two states become possible. If oceanic gravity waves are then introduced along with a seasonal cycle, it is found that any random noise in the model can trigger amplifications of the El Nino events which could grow into the major events which concern us. In other words, Vallis, through the use of a very simple and naive model, by comparison with the very large, physically realistic coupled models now being developed, has been able to examine the broad features of the phenomenon, and has obtained clues which point us to an interesting possibility. This is that the fully interactive and coupled system which is the real world may be exhibiting characteristics of what has popularly become known in recent years as a 'chaotic' system (see, for example, Gleick, 1987), in that switches of state may be triggered by very small amplitude fluctuations, or noise, or that amplification of particular modes may be caused by such triggers of highly nonlinear processes. We do not have space here to go into details of what constitutes a chaotic system, but we might finally note that a chaotic system can exhibit quasi-stable periodicity, and even some degree of predictability. This must surely be a fruitful line of research to develop in the future.

3.3.4   The role of space
The study of the El Nino–Southern Oscillation phenomenon has only become possible in the period since the Second World War, since which time measurements of oceanic and atmospheric parameters have been placed on a relatively firm basis, and statistical evidence for complex processes has been built up. Even so, progress was limited by the sparseness of data over many regions of the globe, and the Pacific Ocean (which obviously is the major player in ENSO and its influence on the world climate) had been one of the most poorly covered regions of the globe prior to the advent of space data. Particularily the southern Pacific had been an almost complete blank in our global data sets.

However, the accurate measurement of global parameters such as ocean surface temperature, surface wind stress, upper air winds, greenhouse gases and even ocean currents is now a possibility from orbiting spacecraft (though the challenges we face in developing instruments and techniques is in some cases fairly daunting). The complex, multi-dimensional, highly interactive models which are now being developed around the world will require such data, both to test that the models are capable of simulating past situations, and to be used as initialization data at the commencement of model prediction experiments. There is no other way of economically gathering the required quantities and accuracy of data other than by space observations. The case does not need further elaboration — it is made by data such as we have already shown in Plate 2. However, we do have to face up to the fact that we are still limited by our ingenuity in what we can actually measure and how accurately. There is a pressing need for the development of novel measurement techniques and instruments to gather all the information which we need to crack the problems of the El Nino and the global environment which face us. For example, we need to be able to measure surface temperatures with higher absolute accuracy; surface winds and waves need an accurate and consistent technique; and the exchange of heat, latent heat and moisture need to be monitored reliably and globally from space. More of this in Chapter 5.

3.4   POSTSCRIPT TO CHAPTER 3

Much research remains to be done on the El Nino phenomenon, the greenhouse effect and the ozone hole and, indeed, on numerous other important processes which we have not been able to highlight. In many cases, conflicting theories and ideas abound, and we may be a long way from an adequate explanation of what is going on. However, in each case we are confronted with problems which require the exploration of the very frontiers of the science of large, nonlinear systems, requiring the application of the best scientific brains which we can bring to bear on the problems. In each case, also, we are confronting problems which carry the most serious implications for the future of mankind on planet Earth. We shall now turn our attention to how we can help to understand and solve these and other environmental problems through the remote observation of our planet on a global basis from space.

4

Space observations — principles

Observations of the Earth from space give us a global view of our planet, and in order to tackle certain problems associated with the climate system, that global view is essential. This is the principal justification, from the point of view of science, for spending the necessary large sums of money to develop, launch and operate Earth observation space systems. There are, of course, other justifications for expenditure on space programmes, deriving from space science and astronomy, communications, resource exploitation, microgravity and defence interests, but it is the global perspective that is the unique benefit of space for climate research. Other advantages that space observations offer include the prospect of repeated observations over long periods using well-calibrated, common instrumentation, and considerable uniformity of sampling over the globe.

In order to be able to study the scientific properties of any part of the climate system from space, we need to be able to 'remotely sense' that component of the system. The information which we have to operate on is basically the spectrum of electromagnetic radiation which arrives at an orbiting spacecraft, having interacted in some way with the Earth's surface and/or atmosphere. As we saw in Chapter 2, this radiation includes the near-ultraviolet and visible, the infrared, and the microwave parts of the spectrum (e.g. see Fig. 2.3), and can be thermally emitted by the atmosphere or surface, or, in the case of solar radiation, transmitted through the atmosphere and scattered by the air or the surface. This applies to the use of natural radiations, such as the thermal radiation emitted by the atmosphere or surface by virtue of its temperature, or the radiation from the Sun. We can also employ 'active' radar or lidar techniques, in which artificial — and usually monochromatic — radiation from microwave (or radio) or laser sources can be directed at the atmosphere or the surface, and the reflected or scattered radiation measured.

The instruments which are to detect and analyse this radiation have to be carried on spacecraft which provide them with a 'home' in the hostile environment of space, supplying the instruments with power, heat, communications with the ground, some protection from hazardous radiations, and so on. The spacecraft need to be placed in orbits which allow a suitable view and coverage of the entire globe, and need to provide a stable and reasonably well known orientation in space, in order for controlled observations to be carried out.

We examined some of the basic principles of radiation transfer, spectroscopy and related topics in Chapter 2. In the present chapter we will consider some of the more practical aspects of observational techniques and the application of some of the basic physics we studied earlier to the real problems of remote sensing of the climate from space. This will include a discussion of some instrumental techniques, aspects of orbital geometry, some details of spacecraft, and a brief look at differing types of observational methods.

4.1 ORBITS

A spacecraft can, in principle, be placed in any one of a number of distinct types of orbit around the Earth, depending on the requirements of the particular application in mind. The variables at our disposal are the altitude, angle of inclination and degree of asymmetry of orbit. Fig. 4.1 illustrates the basic parameters needed to describe an orbit: for our present purposes we will consider just a few of the basic types of orbital configurations which are important for Earth observation.

The polar (or near-polar) orbit is in the general class of Low Earth Orbits (LEO

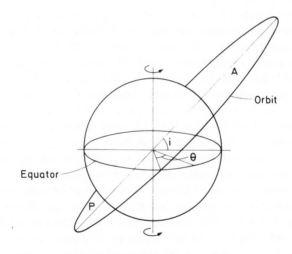

A = Apogee (km)
P = Perigee (km)
i = Angle of Inclination (degrees)
θ = Phase Angle (degrees)

Fig. 4.1 — Illustrating orbital mechanics.

in the customary jargon), that is, at altitudes of a few hundred to a few thousand km above the Earth's surface, with angles of inclination to the equatorial plane of something near 90 degrees (see Fig. 4.1). Once it is in orbit around the Earth, we can imagine the spacecraft following a near-circular path which is fixed — to a first approximation — with respect to an inertial frame of reference: in simpler terms fixed with respect to the stars. Imagine, then, our spacecraft orbiting the Earth in such a way, and it is easy to visualize the Earth rotating on its axis 'below' the spacecraft. As the Earth rotates, and the spacecraft follows its polar orbit, most regions of the Earth's surface will eventually pass under the spacecraft. In other words, we have a way of achieving global coverage. Hence the utility of the polar orbit for Earth observation. By adjustment of altitude, and therefore orbital period and inclination, we can tailor the orbit to suit our exact requirements. One of the most important distinctions at this point is whether the orbit is fixed with respect to the Sun, known as a Sun-synchronous orbit, or whether the phase of the orbit is allowed to drift with respect to the direction to the Sun. Many Earth observation satellites are placed into Sun-synchronous orbits, so that observations are made at the same 'local time' below the spacecraft, at least at each latitude zone. However, if we wish to study a phenomenon that varies strongly with local time (e.g. the photochemistry of ozone in the stratosphere, which depends directly on the intensity of sunlight), we must beware of 'aliasing' the observations, i.e. not obtaining a representative sample of the diurnal cycle, and therefore biasing the data (see, for example, Wunsch, 1989).

The choice of altitude of an orbit in the polar case is a compromise between several factors. First, the lower the altitude then the better the spatial resolution that can be achieved, simply because the spacecraft is nearer the Earth's surface. Second, and on the other hand, a higher altitude affords a better coverage of the surface. Third, at altitudes below a few hundred km the decay of the orbit owing to drag, and the consequent effect on mission lifetime, becomes significant. Moreover, on a typical Earth observation spacecraft the requirements of the different instruments carried on a modern 'observatory' may pull in quite different directions, so that the choice of orbit sometimes becomes a very difficult compromise. Fourth, the lower the orbit, then the stronger are the signal returns that will be received at the spacecraft when using active techniques such as lidars or radars.

At orbital inclinations near 0 degrees we can envisage two useful types of orbit. First, a LEO with inclinations of a few tens of degrees will permit close coverage of the tropical belt of the globe. The main Space Station, 'Freedom', to be put in place by NASA, ESA and other national space agencies, will utilize an equatorial orbit of this sort, largely because this is a relatively easy orbit to access. Second, the equatorial geosynchronous or geostationary (GEO) orbits are enormously useful for Earth observation. These are orbits in which the orbital height is increased (to something like 36 000 km) until the orbital period exactly matches the rotational period of the Earth; in other words, the spacecraft appears to hover above a single point on the equator. The height of this orbit is such that the problem of achieving a high spatial resolution at the surface is severe, whereas the coverage of the Earth is excellent from such a vantage point in space. Not surprisingly, whole-Earth observations are made from such orbits, and global telecommunication satellites also frequent them. Indeed, such is the popularity of geostationary orbits that the problems of overcrowding are getting serious, and require concerted international efforts to solve.

From the geostationary orbit, the development of time-dependent phenomena such as weather systems can be watched with ease, as anyone who has seen a near-real-time display of Meteosat data knows. The development and movement of storms, fronts, clouds and other meteorological systems can be monitored and used to assist in forecasting. The European Meteosat spacecraft system deploys a satellite in a geostationary orbit over the Greenwich meridian, and the USA and Japan (and notionally also the USSR) are responsible for similar spacecraft deployed at other longitudes around the globe, as part of the so-called 'World Weather Watch'. Meteosat is operated by the European agency EUMETSAT, and launched by ESA.

Many other orbital configurations are possible, of course, some of them stable (i.e. repeatable and in phase with a particular geocentric phenomenon). One example is the Molyneux orbit, named after its discoverer, a Soviet scientist. This orbit is highly eccentric, with an inclination of about 63 degrees, which produces a stable orbit with a very large apogee and a small perigee, and which can be arranged to be geosynchronous with the Earth's rotation. This orbit has been considered in a recent design study for an experimental spacecraft mission (T-SAT, 1988), in which an apogee of about 39 000 km, and a perigee of about 1000 km were derived. During the long segment to and from apogee the spacecraft appears to almost hover over a fairly localized region (although its distance above the Earth's surface is varying considerably all the time). Thus, this type of orbit can be used as a high latitude quasi-geostationary (though it is more correctly a geosynchronous) orbit. Because of the rapidly varying spacecraft–surface distance, though, it is perhaps more useful for communications purposes than for remote sensing.

4.2 OBSERVATIONAL MODES

A number of ingenious techniques have been developed to cope with how to address the question of observational geometries for an Earth observation experiment in space. If we are interested in studying the surface of the Earth, the problem reduces to a two-dimensional one, but of course the situation is more complex if we need to probe the atmosphere in the vertical dimension, as we do, as well as horizontally. In this part of the book we will be taking a brief look at some of the better-developed techniques, particularly those which can be used for observations of value in our study of the planet's climate. We will, however, be excluding a number of methods which are still under development, for example the use of active experiments on spacecraft, such as those using laser techniques.

4.2.1 Remote sensing of the surface

Let us first of all deal with the 2-D surface problem. Here we can imagine that the image of a single element of surface (a 'pixel') is imaged by some on-board optics onto the focal plane of a camera , or some other optical or spectroscopic device. The problem is again one of conflicting requirements, which we examined earlier. First, we have to dwell on a single pixel long enough to build up an acceptable signal-to-noise ratio in that pixel. Second, we have to develop some technique for taking in pixels on either side of the subsatellite track, in order to achieve global coverage with the minimum number of orbits: this may require the use of some mechanical or electronic scanning mechanism from side to side, or perhaps the use of arrays of detectors to sweep a wide swath under the satellite. Third, the amount of time available to do any of these things is severely limited by the very high speed over the surface (about 7 km/s) of a polar orbiting spacecraft. As we noted earlier, the height of the spacecraft affects both the spatial resolution achievable and the coverage of the globe that is achievable.

The sweeping or 'pushbroom' technique using multi-element detector arrays benefits from the obvious advantages of a more efficient use of time, and so has been used widely in a number of applications. Alternatives that have been employed include the use of side-to-side scanning mirror assemblies, to image different parts of the scene below the spacecraft onto the detector(s). In the microwave region it is also possible to develop electronically scanning techniques, which can operate with a fixed receiving antenna. A variation on the mechanical side-to-side scanning technique is to employ conical scanning, which has the advantage that a continuous motion of a mirror (to describe a conical acceptance field) is somewhat easier to implement than having to repeatedly reverse the motion of a scanning mirror. The conical scan approach has in fact been used in the design of the Along Track Scanning Radiometer (ATSR), which is to fly on the ERS-1 satellite. The rotation rate of a single rotating mirror in the optical system can be adjusted to correspond to the satellite speed over the surface, so that at each new scan the spacecraft has advanced by exactly the distance corresponding to one pixel. The ATSR will be described further later in this chapter.

A further important technique in surface observation is to use stereoscopic imaging in order to emphasize — and in some cases measure — topography, structures or other features. This can be accomplished for stationary objects by taking images at different times, during different orbits. For objects (e.g. clouds) which move appreciably during the time between successive orbits (about 90 minutes in LEO), a double camera system may be used, though this achieves very little lateral displacement. For intermediate cases, where delays between images of a few to a few tens of minutes can be accepted, conical and other fore–aft scanning techniques can be used.

4.2.2 Remote sounding of the atmosphere (nadir)

If we now turn to the much more difficult problem of remote sounding of the atmosphere in its depth as well as horizontally, we are presented with the problem that there are no recognizable boundaries, as there are when viewing the surface, to use to locate the direction of view. The observation itself must define its own location and boundaries; and techniques must be developed to probe the depth of the atmosphere as well as in the horizontal directions. Here we will consider techniques to do this for the case of viewing the atmosphere vertically below an orbiting spacecraft, that is viewing towards the nadir: the alternative method of viewing towards the limb of the atmosphere will be considered below.

We have already seen in Chapter 2 that the spectral properties of absorption and emission lines in the atmosphere depend on the pressure and amount of absorbing gas in the line of sight. Using Fig. 4.2 we can attempt an explanation of the basic principle of remote sounding. Following the work in Chapter 2, consider a vertical

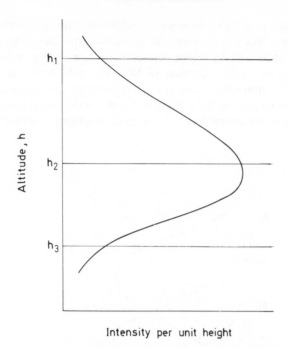

Fig. 4.2 — The principle of remote sounding the depth of the atmosphere.

slice of the atmosphere as shown in the diagram, and three levels in the atmosphere at heights h_1, h_2 and h_3. Consider thermal radiation emitted by each level in the upwards direction (towards our orbiting spacecraft), and let us assume that we have chosen a wavelength (section 2.2.2) on the side of a reasonably intense absorption band of, say, carbon dioxide. At the highest level, h_1, because the air is so thin, relatively little radiation is emitted by that level, but in view of the fact that the air above h_1 is even thinner, this emitted energy suffers rather little attenuation on its journey towards our spacecraft and the sensors on board. Going to the lowest level, h_3, however, here the pressure and amount of CO_2 are much higher, so that the intensity of emitted energy at the chosen wavelength is high: however, h_3 is overlain by many layers of heavily attenuating atmosphere, so that here also relatively little radiation escapes to our spacecraft. Clearly, as long as the conditions just described pertain to levels h_1 and h_3, then there is an intermediate level, h_2, which will generate more signal at the spacecraft, because though emission is less intense than lower in the atmosphere, the attenuation by the atmosphere above h_2 is less. Indeed, the logic tells us that somewhere between h_1 and h_3 the amount of signal per unit height in the atmosphere must peak, as shown by the curve in Fig. 4.2. Of course, the right conditions must be chosen — if the wavelength is extremely highly absorbing, then the atmosphere will be impenetrable at all heights, and our satellite instrument will receive radiation from only the very top of the atmosphere. If, on the other hand, we have chosen an extremely weakly absorbing wavelength, we will receive a small signal from the whole column of the atmosphere, and 'see' right through to the surface.

Fig. 4.2 gives us the vital clue to how we can remotely sound the atmosphere in its depth. If we imagine a series of wavelengths, chosen so that they possess a series of curves like the one shown in the figure (these curves are known as 'weighting' or 'contribution' functions — see equation (2.24)), but with peaks at a variety of different heights in the atmosphere, and if we design our spacecraft instrument to be sensitive to those chosen wavelengths, then each signal will be largely representative of a different height in the atmosphere. A series of weighting functions is shown in

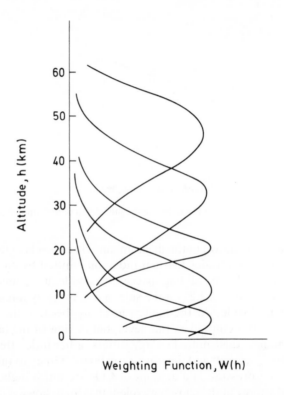

Fig. 4.3 — Illustrating atmospheric weighting functions.

Fig. 4.3: clearly, the technique has a lot more to it than we have described: for instance, the signals received at each different wavelength have contributions from many different levels in the atmosphere, and so are to some extent 'jumbled up'. However, within the constraints of information theory (in simple terms: you don't get information for nothing), techniques of data analysis can be worked out that allow us to get around such problems.

We can state the method in slightly more mathematical terms for those so inclined, by again recalling equation (2.23) from Chapter 2, which gave us the expression for the thermally emitted signal received at the top of the atmosphere:

$$Ev = \int B(v,T(z)) \ d\tau_v \ , \tag{4.1}$$

where $B(v, T(z))$ is the Planck function, and the integration is from the surface to the top of the atmosphere. We can rearrange equation (4.1) to give, for the received intensity,

$$E = \int \int B(v,T(z))W(v,z) \ dz \ dv \ . \tag{4.2}$$

The term:

$$W(v,z) = \frac{\partial \tau_v(z)}{\partial z} \ , \tag{4.3}$$

the first derivative of the transmission with respect to height, is the weighting or contribution function that we encountered in Chapter 2 (equation (2.24)). This term tells us how much each level in the atmosphere contributes to the observed signal at each wavelength. The shape of the weighting function as a function of height is the same basic shape as the curve shown in Fig. 4.2. By knowing the detailed mathematical expression for the weighting functions at all wavelengths used, we are able to quantitatively retrieve the vertical profile of either temperature or gas concentration from the measurements made by our sensor at the various wavelengths we have tuned it to.

4.2.3 Remote sounding of the atmosphere (limb sounding)
There is a fundamental limitation to how small the vertical resolution can be made in the technique just described, basically dictated by the width (in the vertical dimension) of the weighting functions in Figs 4.2 and 4.3. This limit is itself dictated by the exponential dependence of pressure on height in the atmosphere, which means that, no matter how good the spectrometer is that we fly on our spacecraft (i.e. how high a spectral resolution we use), we cannot make the weighting functions sharper, and hence our vertical spatial resolution better. To improve on the vertical resolution, and the fineness of vertical variation which we can distinguish, we must turn to alternative techniques. The most widely used technique today for achieving high vertical resolution is known as 'limb sounding' (Gille and House, 1971). This means that, quite simply, we direct our instrument to look towards the 'limb' of the atmosphere, just above the horizon. By designing the instrument to have a very narrow optical field of view in the vertical direction (we can imagine the instrument subtending a very narrow 'pencil' field on the atmosphere) we can geometrically pick out the layer of the atmosphere which we wish to investigate. Actually, there will be contributions from layers above the chosen layer in this technique also, but we can

envisage that for layers below the bottom edge of the field of view, no radiation whatsoever is received by the instrument (ignoring diffraction effects, which are important in the design of real instruments, but which we will not consider in this book).

We will not develop the mathematics further because they are essentially the same as in the case of nadir sounding, just considered. However, we may imagine the weighting functions of the nadir sounding case to be modified by a multiplying function which represents the additional geometrical 'cut off' below the peak caused by the viewing geometry. On the basis of these ideas, we can represent the limb sounding weighting functions as shown in Fig. 4.4, which quite clearly shows the enhancement of vertical resolution over the nadir sounding case, due to the sharp cut-off at the lower edge of the field of view.

The technique possesses another considerable advantage for detecting and

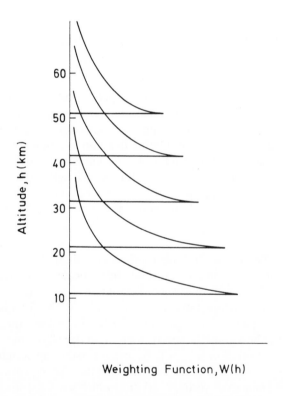

Fig. 4.4 — Weighting functions in the case of limb sounding.

measuring very weakly radiating trace gases in the upper atmosphere, which is that by taking a 'sideways' look at the atmosphere, we are effectively integrating along a very long atmospheric path, and so increasing the strength of the signal which we can receive. This is offset in the lower atmosphere by a certain disadvantage, caused by that same long path in the atmosphere: it is difficult to use the technique down into the troposphere, because the existence of clouds and general haze means that over such long paths the overall level of attenuation becomes very high. For these two reasons, the method of limb sounding finds its greatest application to observations of the stratosphere and mesosphere.

4.3 INSTRUMENTAL PRINCIPLES

Fig. 4.5 gives a 'family tree' of spectroscopic instrumental techniques for remote

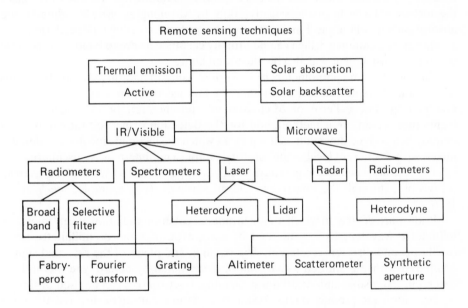

Fig. 4.5 — A 'family tree' of remote sensing techniques and instruments.

sensing from space. Since a complete development of the descriptions of these techniques is worthy of a book itself (for example, note the review in Harries, 1982a), we can only refer here to some of the broad principles which will be illustrated by reference to just a few of the methods shown in the figure. At the top of the diagram we see four of the basic observational methods available to us, classified by the nature of the radiation source and observing geometry. Below this we have divided the classification into visible and infrared methods on the one hand, and microwave methods on the other. The three subclasses under visible/infrared depend on spectral resolution, beginning with fairly broad band radiometric techniques at left, and going through 'classical' high dispersion spectrometers, to monochromatic laser techniques. Under microwaves, we define two broad classes: radar, and heterodyne radiometry.

Heterodyne systems employ a monochromatic local oscillator which is mixed in a nonlinear element with the signal, to produce a signal at a difference frequency which is more amenable to subsequent analysis than the original frequency. We will illustrate an example of microwave heterodyne spectroscopy later. Laser techniques can be used either in a 'passive' mode as a local oscillator in a heterodyne system, or if the laser is sufficiently powerful, as a backscattering lidar. More conventional high dispersion techniques are, perhaps, better known, for example the grating spectrometer or the Fourier Transform Spectrometer (FTS); one of the examples of instruments that follow will be based on a grating spectrometer. Selective filters are that class of device in which high selectivity is achieved by including within the instrument itself a sample of the gas to be observed; by arranging a suitable means of modulating the transmission of the gas (e.g. by modulating the pressure of the gas), the instrument can be made sensitive only to wavelengths corresponding to the absorption lines of the gas. Finally, if only low spectral resolution is needed, there are a number of techniques which can be broadly classified as broad band radiometers; this type of instrument will also be illustrated later.

Radiometers and photometers isolate broad bands of natural radiation by means of spectral filters of one sort or another; by gratings or other varieties of spectrometers (e.g. Fabry–Perot or Michelson interferometers); or by selective filter techniques. A different method, still for the detection of natural radiation, is to employ heterodyne detection techniques, in which the incoming incoherent signal is mixed with a coherent local oscillator, for example a millimetre wave oscillator, or a laser. Typically, such techniques have been developed and used at wavelengths from the visible, through the infrared to the microwaves. (For example, see Waters, 1976.)

Radars are devices in which monochromatic, coherent radiation at centimetre or millimetre wavelengths is generated by some type of radio or millimetre wave oscillator, and directed at a target. The strength and phase of the returned signal, compared with the emitted signal, can be determined by heterodyne methods of detection. At visible and infrared wavelengths, laser sources can be used in similar configurations, to produce intense beams of radiation that can be directed at targets such as the atmosphere or the surface, and the return signals detected in similar ways by mixing in nonlinear crystals with 'local oscillator' radiation derived from the same laser source. (For example, see Ulaby, Moore and Fung, Vol. II, 1982, and Vol. III, 1986.)

In what follows, we will describe four quite distinct techniques, for illustration. These are an infrared limb sounding radiometer, a microwave radiometer, a visible/infrared imager, and a radar altimeter. It is stressed, however, that the profusion of subtly different sensing techniques and instruments is very diverse, only being limited by the ingenuity of scientists in their creation, as the complexity of Fig. 4.5 will demonstrate.

4.3.1 Infrared limb sounding radiometer

We have already seen in Chapter 2 that the intensity of thermally emitted radiation carries information about the temperature and thermal emissivity of the emitting body or surface. We saw there that the gases in the atmosphere have complex spectra, containing large numbers of individual spectral lines of various intensities, but all possessing a similar basic spectral shape (see Fig. 2.5). The spectral properties of the land surface show much broader features, but these can also often be associated with specific minerals or vegetation cover. The oceans, and other water masses in general, show only very slowly changing spectral properties as a function of wavelength. These differences arise because the molecules of the gases that make up the atmosphere are essentially separate and uncoupled, whereas in a liquid or a solid the closer proximity of molecules broadens or blurs out spectral features.

In the visible and the shortwave infrared, where the intensity of solar radiation is high, the spectral properties of the atmosphere and surface can also show up in absorption of scattered or reflected light at specific wavelengths characteristic of the molecules present. For many applications, we do not need to isolate a single spectral line in the atmosphere, but we can increase the detected signal-to-noise ratio by detecting a number of spectral lines of the same gas, integrating over perhaps a hundred wavenumbers or so; another example might be in studying the mineral properties of surface rock, where again we need to isolate only fairly broad spectral bands characteristic of certain minerals.

Thus, in this case the problem reduces to one of first isolating the selected spectral band(s) of interest, and second then detecting the isolated radiation with a suitable detector. As an example of how this problem might be solved, we will describe an infrared radiometer which flew on the Nimbus 7 spacecraft. The experiment was called the Limb Infrared Monitor of the Stratosphere (LIMS) (Gille and Russell, 1984). LIMS was designed to make measurements of the temperature and composition of the stratosphere, by isolating specific bands in the infrared where radiation could be detected from a number of gases. CO_2, emitting radiation at 15 microns, was to be used to measure temperature, because the concentration of CO_2 in the atmosphere was known to be well mixed at about 330 ppm: thus the intensity of infrared radiation detected by an instrument mounted on an orbiting spacecraft could be related to variations in the temperature of the CO_2 (and therefore of the atmosphere) because the concentration was known. Measurements at other wavelengths corresponding to absorption/emission bands of ozone, O_3, nitric acid, HNO_3, nitrogen dioxide, NO_2, and water vapour, H_2O, could then be related to the concentrations of these gases, because the temperature was known from the CO_2 channel.

Table 4.1 gives the details of the channels selected for LIMS. The problem of isolating the radiation as required from the various channels was solved by using interference filters, which had been developed for a variety of infrared applications over the preceding 20 years or so. The operation of these filters is in principle very simple. They consist of alternate layers of two different dielectric materials with differing refractive indices at the infrared wavelengths concerned. Suitable materials might be germanium, silicon, indium antimonide, indium phosphide, and similar semiconductors. The thickness and arrangement of the layers is designed so that the stack acts as a multiple Fabry–Perot interferometer (Born and Wolf, 1975): a series of regular pass bands is first designed into the filter, and other thicknesses are then incorporated so as to isolate just one of these pass bands, at the wavelength desired. The beauty of such filters is that, within limits, they can be tailor-made to match specific requirements: the skill of the designer and manufacturer is to arrange the required spectral selection, while achieving high out-of-band rejection (no 'stray light'), and good transmission within band.

Table 4.1 — Details of LIMS channels

Channel	Emitting gas	5% band-pass points (cm^{-1})	Vertical FOV^a (km)	Horizontal FOV (km)	Noise equivalent radiance (W/m^2 sr)
1	NO_2	1560–1630	3.6	28	0.00055
2	H_2O	1370–1560	3.6	28	0.0023
3	O_3	926–1141	1.8	18	0.0037
4	HNO_3	844–917	1.8	18	0.0015
5	CO_2 (wide)	579–755	1.8	18	0.0055
6	CO_2 (narrow)	637–673	1.8	18	0.0014

^{a}Field of view.

The detectors employed on LIMS were also based on semiconductor materials, in this case mercury–cadmium–telluride (M–C–T) amalgams which behaved as photo-conductive detectors. These operate by absorbing a suitably energetic photon (the low energy cut-off for these materials typically corresponds to mid-infrared wavelengths between 10 and 25 microns) which excites an electron from the valence band of the material, into the conduction band, where it is detected as a current by applying a suitable bias voltage. Such detectors are swamped by thermally excited electrons at room temperature, and so for sensitive operation they need to be cooled, to temperatures below 100 K in this case. This, of course, presents a problem for a space experiment where weight is at a premium, but was achieved in LIMS by flying a solid cryogenic cooling system comprising a combination of solid argon and methane. This allowed the detectors to be operated at temperatures of about 65 K, but meant that the lifetime of the experiment was limited to about 7 months, which was the design life of the solid cryogen. The cryogen was being used up continuously in orbit, subliming directly from solid to gas as it absorbed the heat of the detectors and surroundings.

More details of the principles behind the interference filters and the photoconductive detectors used on LIMS are available in a variety of sources, for example Houghton and Smith (1966).

Using interference filters to define the spectral pass bands of the instrument (Fig. 4.6), and the M–C–T detectors cooled as described to detect the infrared radiation, the overall configuration of the LIMS experiment was as shown in Fig. 4.7. The input scan mirror could be moved to scan the so-called 'limb' of the atmosphere, just above the horizon, in the observational mode known as limb sounding (see above). The incoming beam of infrared radiation was then divided and directed through a set of interference filters, to fall onto a focal plane array of detectors. The radiation was 'chopped' by a mechanical chopper to produce alternating signals in the detectors, which were then amplified by a cooled preamplifier before being passed through a series of electronic processing stages and delivered to the spacecraft's electronic system for recording and for transmitting to the ground.

Nimbus 7 was placed in a near-polar orbit (see elsewhere in this chapter for

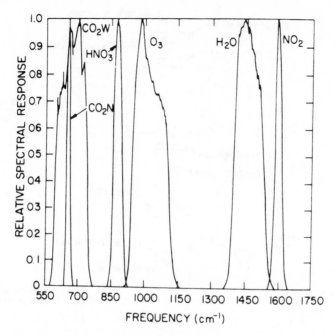

Fig. 4.6 — The spectral response (transmission) of interference filters used in the LIMS experiment.

spacecraft and orbit details) and LIMS viewed to one side of the spacecraft: the limb of the atmosphere was therefore 'swept' by the field of view of the instrument, and a global coverage built up in successive orbits of the spacecraft. The signals in each spectral band were reconstituted on the ground from the transmitted data stream, and the temperature and composition of the atmosphere calculated from these signals. In order to do this, the equations developed in section 2.1 earlier were used as a basis. We can use the equations relating the signal received in a particular spectral channel to the temperature and density of the relevant gas, averaged along the atmospheric line of sight, as given earlier in equations (4.2) and (4.3).

The term $B(v, T)$ in these equations is, of course, the Planck function, which is proportional to atmospheric temperature (see equation (2.12)); and the term $\partial\tau/\partial z$ represents the first derivative of the transmissivity of the atmosphere in the selected band: this obviously depends on how much gas there is in the line of sight (the density of gas integrated over the line of sight — see section 2.1.7). In equation (4.2) the extra integration over spectral frequency, v, is to account for the width and shape of the filter pass band.

In the case of CO_2, we know the atmospheric concentration of the gas, and so we can derive the temperature of the atmosphere from a solution of equation (4.2). In the case of gases of unknown concentration, we can then utilise the temperature so calculated to derive the $\partial\tau/\partial z$ term, and thence the concentration. Of course, these simple statements hide the fact that there is a great deal of complexity and subtlety

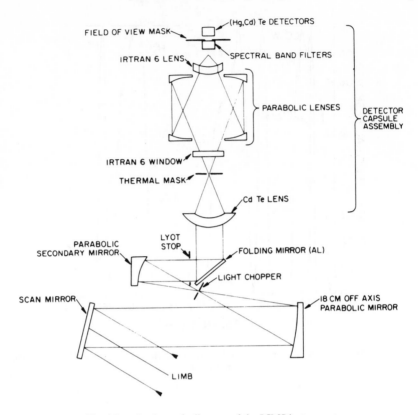

Fig. 4.7 — A schematic diagram of the LIMS instrument.

involved in using equations (4.2) and (4.3) in this way. For example, the terms in the equation are complex and nonlinear (that is, one parameter is not related to the others in a simple multiplicative way). Also, it has to be remembered that radiation in any one pass band of the instrument is arriving at the instrument from a wide range of points along the limb path, and this means from a range of effective altitudes. For these and other reasons the analysis of the radiometric signals, or the 'retrieval' of data as it is called, is far from simple, and has given rise to a considerable number of papers in the literature (for example, see Houghton *et al.* (1984) for further references) on mathematical techniques to extract the maximum amount of data available, with a minimum of distortion from noise. In Chapter 5 we will show some of the results that LIMS achieved during its 7-month lifetime on the global distribution of a number of stratospheric trace gases, some for the very first time.

4.3.2 Microwave radiometry
The microwave region of the spectrum is characterized by a much higher transparency of the atmosphere compared with the high absorption properties of the infrared. Not only is the molecular absorption weaker than in the infrared, but the effect of scattering is itself much weaker at these longer wavelengths (see section

2.1.7.2). For this reason, the microwave region is of great value in making observations from space of not only the atmosphere but also the surface. The main features of the spectrum were shown in Fig. 2.4. The major molecular absorbers in this region are water vapour, H_2O, and oxygen, O_2, though a number of other gases, amongst them ozone, O_3, and others such as N_2O, CO, HNO_3, and ClO, also have measurable absorption/emission lines. One significant advantage of this region of the spectrum is that heterodyne techniques can be used to detect very weak signals, and also to operate at very high spectral resolution, since the local oscillators that are used in the technique can be designed to have a very narrow bandwidth. Also, because of this property of very high spectral resolution, it is possible to make measurements of the slight shifts of spectral lines due to the Doppler effect arising from the motion of air in the atmosphere; i.e. we have a means of making direct measurements of wind.

The details of the spectrum are, of course, considerably more complex than

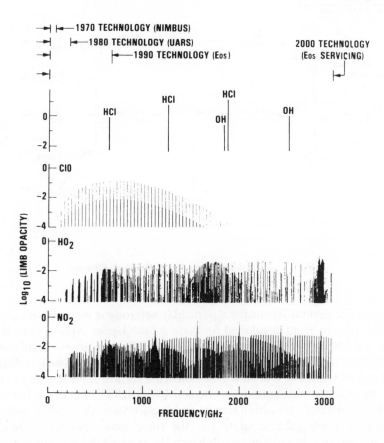

Fig. 4.8 — Details of the microwave spectrum of the atmosphere.

shown in Fig. 2.4, and to illustrate this we can show Fig. 4.8, taken from the work of Waters (1989). This shows line spectra of several other gases, including NO_2, ClO, HO_2, HCl, and OH, and the relative intensities of the individual lines, right up to very high microwave frequencies of about 3000 GHz (wavenumber, $v = 100 \, cm^{-1}$). Also shown in this diagram are the approximate limits imposed by current technology at different times. For instance, at the present time, US and UK teams are working to fly a system operating up to 600 GHz or so on the Space Station polar platform. Advances in this field are largely determined by the ability to make smaller and smaller components, corresponding to higher and higher frequencies, and at the same time to keep internal noise, stray capacitance, and other features which limit performance, to a minimum.

The basic technique can be explained by reference to Fig. (4.9) which shows in

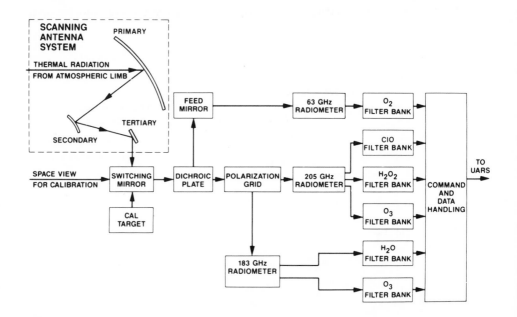

Fig. 4.9 — The design of the Microwave Limb Sounder.

block form the general layout for a particular heterodyne microwave radiometer, this one currently being developed for flight on the Upper Atmosphere Research Satellite (UARS) in 1991 (Waters, 1989). The diagram shows radiation being received from the atmosphere via a scanning antenna system, which is diffraction-limited and subtends a field of view of 3.5 km in the vertical at the distance of the limb of the atmosphere. Calibration is effected by switching the input between the atmosphere, internal calibration targets and a space view. A series of dichroic and polarizing elements then separate out the three main spectral channels. The signals are detected in radiometers in which the sky signals and signals from local oscillators at 63, 183 and 205 GHz, are mixed in Schottky barrier-type mixers, and down-converted in frequency to intermediate frequencies (IF) in the range 0 — 3 GHz. These IF signals are then amplified and sorted into six bands, each centred at 400 MHz and with 500 MHz bandwidth, before being passed into six filter banks, which each split the signal into 15 individual spectral channels. The power in each of these channels is then measured and the value digitized for transmission to the ground.

This instrument, known as the Microwave Limb Sounder, MLS, comprises three principal components: the sensor assembly, the spectrometer, and the power supply. It weighs about 290 kg and requires about 185 W of power; it produces data at the output at a rate of about 1.25 kb/s. A sketch of the antenna and the radiometer box is given in Fig. (4.10), which can be scaled if it is remembered that the long axis of the primary antenna measures 1.6 m.

The next generation of microwave sounder will be the EOS MLS, to fly on the

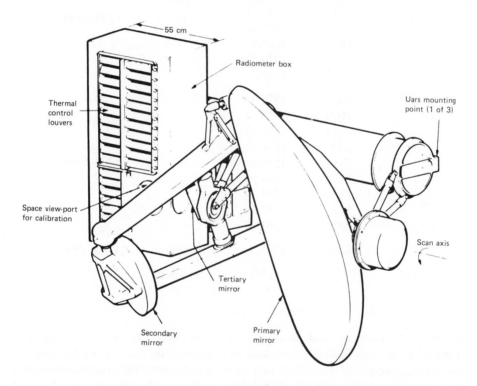

Fig. 4.10 — A schematic view of the Microwave Limb Sounder.

NASA EOS (Earth Observing System) polar platform, at the end of the 1990s. This will extend the design of the UARS instrument, principally to achieve higher performance and higher spectral frequencies. Table 4.2 records the primary channels and measurement goals for the EOS MLS, from which it will be seen that channels at 560 and 637 GHz are to be included. It is worthy of note that the 560 GHz channel is at the heart of the hardware to be provided by the UK in this project. Both the 560

Table 4.2 — EOS MLS channels

Basic parameters	Mass	450 kg
	Power	650 W (average)
	Data rate	1 Mb/s
	Size	$1.5 \times 2.0 \times 1.2 \text{ m}^3$

Channel (GHz)	Species (frequency) (GHz)		Operating temperature (K)	Radiometer noise temperature (K)
637	ClO	(649.5)		
	H_2CO	(647.1)		
	HNO_3	(644.3)		
	CH_3Cl	(637.0)	80	5200–5900
	HOCl	(635.9)		
	HCl	(625.9)		
	O_3	(623.7)		
	HCN	(620.3)		
560	NO_2	(578.4)		
	HO_2	(578.2)		
	N_2O	(577.6)		
	CO	(576.3)		
	BrO	(576.3)	80	5200–6000
	ClO	(574.6)		
	H_2O_2	(561.1)		
	SO_2	(555.7)		
	NO	(551.2)		
205	ClO	(204.4)	Ambient	1700–1900
117	O_2	(118)	Ambient	700

and 637 GHz channels will possibly need to be cooled in order to achieve the required sensitivity, and it is intended that this will be done using Stirling cycle coolers, developed for space flight in the UK. A diagram of the essential elements of the high frequency part of EOS MLS is shown in the next figure, Fig. 4.11, where the 'cross' symbols represent the high frequency Schottky barrier mixers being developed in the UK and the USA for the 560 and 637 GHz channels. These pieces of novel technology are state of the art, that is, right at the forefront of technological achievement worldwide.

The signal retrieval method to be used to derive geophysical parameters from the raw data are based on the mathematical techniques developed in Chapter 2, and also on the limb sounding method described earlier in this chapter. Following on from the previous section, we can write down a simple basic expression to describe the signal being detected by a microwave radiometer of the type we have considered. An

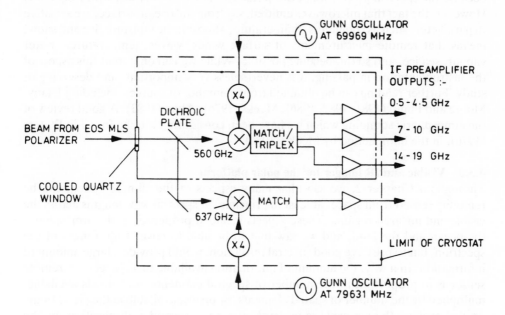

Fig. 4.11 — The high frequency channels of the EOS Microwave Limb Sounder.

important simplification is that, at the very long wavelengths which occur in this spectral region, and for the typical temperatures of emission which occur in the Earth's atmosphere, the Planck function can be simplified , and becomes linearly proportional to temperature,

$$B(\nu, T) = 2kT\nu^2 \ ,$$ (4.4)

and so the equation (4.2) for the radiation received at an orbiting spacecraft can be expressed in terms of the 'brightness temperature' (which itself is equal to the physical temperature along the ray path, multiplied by the effective emmisivity), as in

$$T_{B,\nu} = \int T(z) \, \frac{\partial \tau(z)}{\partial z} \, \mathrm{d}z \ .$$ (4.5)

Microwave sounding can be used not only for observing the atmosphere, of course, but also for studies of the land and ocean surfaces. For such applications, similar techniques may be used, although in these cases the very high spectral resolution is usually not such an advantage as in the atmospheric case, simply because the spectral properties of the surface do not vary very fast with frequency. However, the fact that microwaves emitted, say from the ocean surface, are sensitive to parameters such as roughness, temperature, atmospheric moisture, ice and snow, means that remote measurements of surface winds, waves, temperatures, water vapour, sea ice, and so on, become possible. While we can only treat this aspect of the instrumentation in passing, it is nevertheless very important, and deserving of study. Further reading can be obtained from a number of sources, including Ulaby, Moore and Fung, (1981, 1982, 1986); Meeks (1976); Allan (1983). A good review of the atmospheric sounding which we have been considering is to be found in Waters (1976), in the book edited by Meeks (1976).

4.3.3 Visible and IR imager for the polar platform

Throughout Chapter 2, we saw that many aspects of the climate system can be remotely sensed from space, at least in principle, at spectral wavelengths across the visible and infrared regions. These aspects included properties of the atmosphere, the ocean and the land, and we saw that to be able to cover wide ranges of the spectrum, but to preserve good spectral resolution, would provide a large amount of information. In a way, the amount of information available to a spaceborne remote sensor is in proportion to the number of spectral elements or channels available, multiplied by the number of spatial elements, or pixels, available in the view. In the case of probing the atmosphere in depth, there is a second multiplication, by the number of samples in the vertical dimension. All this is very obvious, though it becomes harder to evaluate the information content, perhaps, when we come to consider noise superposed on the measurements. In that case, we need to be able to design our measuring system using information theory to provide quantitative measures of system performance.

Nevertheless, we can easily appreciate that the higher the number of spectral channels, and the higher the number of horizontal, and if appropriate vertical, spatial elements, then the more information there will be available. This has led to consideration in a number of centres around the world of the design of the 'ultimate' optical/IR sensor, which possesses a large number of spectral channels, a high spatial resolution, and the ability to sound in depth with high vertical resolution. Some attempts have been made to design this ultimate machine, but the task is somewhat impractical, leading to very large, heavy, power-hungry and high-data-rate instruments. Thus, a certain degree of limiting the specification is necessary. In plans for the NASA Earth Observing System (EOS) polar platforms, several instruments have therefore been put forward as candidates for flight, with design specifications targeted at a number of different applications. In this section we are going to illustrate the principles behind the design of one of these new sensors, the Moderate Resolution Imaging Spectrometer, MODIS. This is designed to provide up to 64 simultaneously observed spectral channels, at wavelengths of interest in the study of a number of phenomena important to the climate system, which require only moderate surface spatial resolution.

The basic concept behind MODIS is to use an array of 64×64 detectors in the focal plane of a grating spectrometer. One axis of the array represents variations of wavelength as defined by the grating spectrometer; the other axis of the array represents distance along track (i.e. in the direction of flight). Scanning at right angles to this, in the across-track direction, is achieved by means of a scanning mirror. The concept is illustrated in Fig. 4.12 which shows a schematic representation

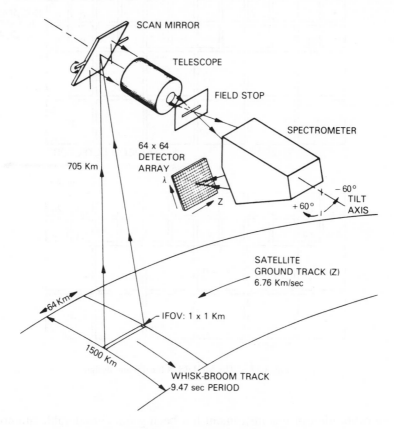

Fig. 4.12 — The MODIS-T scan geometry and system layout.

of the instrument in flight. There are two versions of the instrument proposed, one with a 'tilt' capability to achieve a bigger swath, known as MODIS-T, and one for nadir viewing, MODIS-N. The schematic shown is actually for MODIS-T.

Fig. 4.13 shows the focal plane design for MODIS-N. There are two sets of detectors, one operating at ambient temperature, and the other (for the infrared) operating at 80 K. The smaller detectors subtend a resolution at the surface of 0.5 km; the larger ones subtend a resolution element (pixel size) of double this, that is, four times the area. Table 4.3 illustrates the initial channel selection for MODIS-N, and shows what the various channels will be used for. It can be seen that this instrument will provide data on a large number of the climate problems which we considered in Chapters 2 and 3. The data rate from MODIS-N with 12 channels having 0.5 km resolution, and 24 channels having 1 km resolution, will be 7.6 Mb/s during the day and 1.2 Mb/s at night (when most of the visible channels will, of course, not be used).

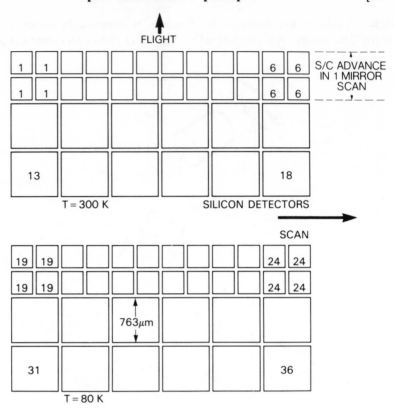

Fig. 4.13 — The MODIS-N focal plane design.

The calibration of this instrument has been given considerable attention. The system will carry on-board visible and infrared calibration targets, although the design of a thermal infrared black-body target for MODIS-N, which is currently specified to have a huge aperture of some 40 cm, has yet to be settled. Space views will, of course, be used to provide a zero or cold calibration point, and measurements will also be made of well-characterized targets on the Earth's surface (for example, White Sands, New Mexico), or by observing the Moon. It is also planned to carry out periodic calibrations using precision calibration systems carried on the Space Shuttle, although this will have to be in a similar polar orbit to the polar platforms of the EOS system. It is foreseen that MODIS will observe continuously and globally, and will include a significant amount of on-board intelligence to control its operating sequences. The instrument will operate as a facility instrument, available to the wide scientific community; it should make a major contribution to the global database for climate research.

4.3.4 Radar altimetry

Radar techniques generally are used to measure the range of a target, by measuring the time delay in detecting the reflection of a pulse of radio frequency or microwave

Table 4.3 — MODIS-N visible/near-IR/IR channels

Channel	Wavelength (nm)	Width (nm)	IFOV[a] (m)	Surface reflectance (%)	Comments
1	470	20	500	3	Soil/vegetation differentiation
2	550	20	500	10	Green peak chlorophyll
3	670	20	500	6.5	Chlorophyll absorption
4	710	20	500	9	Red-NIR transition
5	880	20	500	25	Vegetation max. reflectance
6	960	20	500	24	H_2O peak
7	435	10	1000	5.1	Low chlorophyll
8	490	10	1000	3.5	Non-linear chlorophyll
9	520	10	1000	2.8	High chlorophyll
10	565	10	1000	1.8	Chlorophyll baseline
11	590	10	1000	0.6	Sediment
12	665	10	1000	0.17	Atmosphere/sediment
13	765	10	1000	0.1	Atmosphere correction
14	865	10	1000	0.1	Atmosphere correction
15	754	1.2	1000	30	Cloud altitude
16	761	1.2	1000	90	Cloud altitude
17	763	1.2	1000	50	Cloud altitude
18	500	100	1000	2.5	Polarization
19	500	100	1000	2.5	Polarization
20	1 080	20	500	25	Leaf morphology
21	1 131	20	500	10	Cloud H_2O absorption
22	1 240	20	500	10	Leaf H_2O absorption
23	1 550	20	500	14	Leaf H_2O absorption
24	1 640	20	500	10	Snow/cloud differentiation
25	2 130	20	500	10	Cloud penetration
				NEDT[b] (K @ 270 K)	
26	3 750	90	1000	0.14	Cloud and surface temperature
27	3 959	50	1000	0.14	Cloud and surface temperature
28	4 050	50	1000	0.13	Cloud and surface temperature
30	8 550	500	1000	0.01	Stratospheric aerosol detection
33	10 450	500	1000	0.01	Stratospheric aerosol detection
34	11 030	500	1000	0.01	Cloud and surface temperature
35	12 020	500	1000	0.02	Cloud and surface temperature

[a]Instantaneous field of view.
[b]Noise equivalent detector temperature in Kelvins.

radiation from the target. Radar, because it is capable of very high frequency discrimination or resolution, can also be used to measure the Doppler shift of targets, and therefore their speed in the direction of the line of sight. More advanced radar techniques, for example synthetic aperture or real aperture radars, can provide actual images of the reflecting object or surface, because both the amplitude and phase of the signal can be preserved in the radar technique. The application of radar is largely weather-insensitive, at least except for the heaviest and most intense cases of precipitation, and so is highly valuable for all-weather applications. However, this property of very low scattering coefficients from air or cloud particles also leads to a disadvantage, that radar techniques are generally of little use in the study of the atmosphere, unless they are very powerful, and usually require the presence of a solid or liquid surface to generate significant return signals.

Radar techniques have been found to be of use in a variety of geophysical studies, for example the imaging of surface waves on the ocean, the measurement of surface winds through their effect on surface waves, the imaging of ice-covered regions, and so on. Only some of these have been of use in climate research, however, which generally requires quantitative rather than qualitative image information. Nevertheless, measurements of surface winds and waves by non-imaging scatterometer radars have been shown to be of great promise in studying the air–sea interaction, for example; and the subject of this section, radar altimetry has been shown to be an exciting and promising method for studying global ocean currents, obviously of great importance to the flow of heat energy around the globe.

Radar altimetry from space comprises a narrow beam radar on a spacecraft pointing vertically down towards the Earth's surface. Consider initially the case of measurements over the ocean. A very short pulse of radar energy is transmitted, filling a narrow beam angle, as shown in Fig. 4.14. If the ocean surface is absolutely

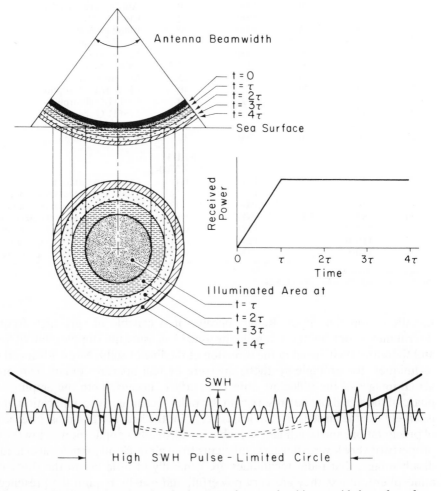

Fig. 4.14 — The interaction of a pulse of radar energy from a radar altimeter with the surface of the ocean see (see text).

flat, the wavefront of the pulse will interact with an area of the surface that spreads out from the central point as shown. The area of ocean illuminated by the radar pulse increases linearly to a maximum, and will then stay constant (from geometrical considerations) as the annulus formed by the beam spreads out further still. If a very accurate clock is included on board, the time taken for the pulse to travel to the surface and back can be determined, and after a few corrections for atmospheric and ionospheric delay, and other aspects, this can be converted to the distance between the satellite and the surface.

Why go to all this bother? Because, if, by the method just described, we can measure the global topography of the oceans, then this can be related to the surface slope of the ocean caused by flowing currents: in other words, the measurement, if carried out with sufficient accuracy, can be a means of deriving the global ocean circulation, of crucial importance to the climate system (see Wunsch and Gaposch-kin, 1980). There is one other major factor to consider, however. The surface topography is indeed related to currents, but additionally is influenced strongly by the geoid, that is, the surface of constant gravitational potential around the globe. This is not a simple sphere or even a simple ellipsoid, because the mass distribution within the solid Earth is not uniform and symmetric. If the geoid can be determined independently, for example from accurate laser tracking of satellites from ground stations around the globe, then the effect of the geoid on the surface topography can be eliminated from the measurements by the altimeter, and the effect of the currents can be isolated.

The effect of the flowing currents of water on the topography may be understood by referring back to equation (2.53) in Chapter 2, which gave the balance of forces acting on moving water at the surface of the ocean:

$$\frac{d\mathbf{v}}{dt} = -\frac{1}{\rho}\nabla P + f\mathbf{v}\wedge\mathbf{k} + \mathbf{F} . \tag{4.6}$$

If we assume for the sake of argument that the conditions of geostrophic flow exist, i.e. that frictional term, $\mathbf{F}$, can for the moment be ignored, then we can see that the horizontal pressure gradient force, $(-1/\rho)\nabla P$, is balanced by the Coriolis force, $f\mathbf{v}\wedge\mathbf{k}$, and the velocity vector is directed at right angles to the pressure gradient. In the ocean, as indeed in the atmosphere, pressure gradients are due to differences in the weight of matter — in this case, ocean — at different points on the surface: in other words, to the head of water between those two points, which is related to the difference in the elevation of the surface at the two points, relative to the geoid. Thus the link between surface topography and currents is established. To quantify the effect: the strongest currents, for example the western boundary currents such as the Gulf Stream or the Kuroshio off Japan, will exhibit a height difference of about 1 m over a horizontal distance of about 100 km — not a huge effect!

Let us return to the operation of the altimeter as an instrument, however. So far, we have considered measurements of the ocean topography only. Our earlier treatment of how the altimeter performs was based on a perfectly flat ocean surface. If we consider the performance in the presence of waves, we find that the altimeter also provides us with a means of measuring the average height of waves in the pulse 'footprint', on a global basis. Such information is of value to climate researchers when studying the interactions of the atmosphere and ocean. Going back to Fig. 4.14, the lower part of the diagram shows what happens to the altimeter pulse when it intersects a surface with waves. The term 'significant wave height' is a measure of the mean height of waves on the ocean surface (SWH specifically means the trough-to-peak height of the highest 1/3 of waves observed). As the altimeter pulse reaches the surface, the effect of the waves is to reflect some signal before the pulse would have reached a flat surface, and to reflect some energy again later than would have been received from a flat surface. The influence of waves, therefore, is to broaden and to skew the return pulse. This is demonstrated by Fig. 4.15, which shows examples of

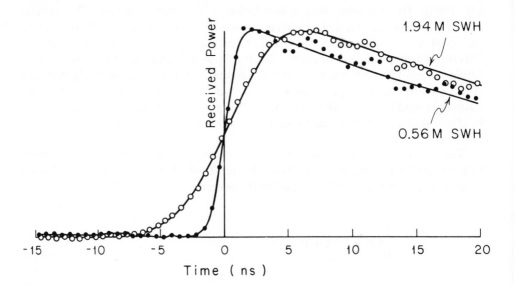

Fig. 4.15 — The dependence of the returned altimeter pulse shape on the significant wave height (SWH).

return pulse shapes, for two different values of SWH. The time = zero point is chosen to be the point at which the first signal return would have been received from a perfectly flat surface. The shape of the return pulse is obviously related to the SWH, and research has shown that this relationship is good enough to use the altimeter pulse shape results to determine SWH on a global basis.

Thus we have seen that the altimeter shows great promise as a tool for obtaining global ocean surface data on currents and on wave height. Not only this, but it is thought that the absolute reflectivity of the signal can also be used to measure surface wind speed, from the effect of surface waves. However, we have only intimated at the difficulties of these various measurements, which are very considerable indeed (though not beyond the realms of possibility). Before we leave the subject of altimetry, we will briefly give the reader some quantitative feel for just how severe these difficulties really are. We have noted in passing that surface gradients of about 1 m in 100 km are typical of the strongest currents. Therefore altimetric topography errors of less than about 10 cm over distances of 100 km are desirable. Table 4.4

Table 4.4 — Typical errors encountered in satellite radar altimetry (taken from Stewart, 1985)

Source of error	SEASAT (uncorrected) (cm)	SEASAT (corrected) (cm)	Future (corrected) (cm)	Wavelength (km)
Geoid	100 (m)	100–200	10–50	200–40 000
Orbits	5 (km)	100–200	5–10	10 000
Coordinate system	100–200	100–200	10	10 000
Ionosphere	0.2–20	0.2–5	1.3	20–10 000
Mass of air	230	0.7	0.7	1000
Water vapour	6–30	2	1.2	50–1000
Electromagnetic bias	4	2	2	100–1000
Altimeter noise	5	5	2	6–20
Altimeter tracker	10	4	0	100–1000
Altimeter calibration	50	5–10	2	∞
Altimeter timing	0	5	0	20 000

shows estimates of the errors involved in the SEASAT altimeter experiment, flown in 1978. The errors shown represent the uncertainty in knowledge of the geoid, of the satellite orbit, of the coordinate reference frame used, and of the delay due to absorption by electrons in the ionosphere, and water vapour in the atmosphere, and other altimetric instrument parameters. Also shown are the distance scales (wavelength) over which these errors apply. The table indicates the errors for SEASAT after the most accurate available corrections were applied, and the likely errors which will apply to future dedicated altimetric satellite experiments, for which the choice of orbits and instrumental techniques will be made to minimize errors. Clearly, many of these error sources are very considerable in size, compared to the signal being sought: it is anticipated that in future missions, it will be uncertainty in the geoid, or reference gravitational field, that will dominate the error budget.

4.4 SPACECRAFT

It is somewhat surprising to recall that spacecraft in Earth orbit have been a reality for only some 30 years, since the time of Sputnik I in 1957. In the intervening period, standards of spacecraft engineering and technology have improved enormously. A spacecraft must be completely self-contained and needs to include systems to provide power, protection against extremes of heat and cold, pointing stabilization, data transmission and command reception. Table 4.5 provides an overview of a number of the spacecraft systems that have been developed over this fairly short period, some of which will now be described in some detail. A very successful series of spacecraft used by the United States for a number of Earth observation missions has been the advanced TIROS spacecraft, shown in Fig. 4.16. This is a 3-axis stabilized spacecraft which provides a mounting plane for a number of remote sensing experiments. One end of the spacecraft contains the basic spacecraft 'bus' systems of power, electrical systems, thermal balance systems, telecommand and so on. The power is provided by solar cells mounted on a solar 'paddle'. This basic spacecraft has been used for the TIROS meteorological operational satellites operated by the US weather agency NOAA.

The Nimbus spacecraft, shown in Fig. 4.17, was used for a number of atmospheric and oceanographic research missions between 1964 and 1978, when the last in the series, Nimbus-7, was launched. Nimbus-7 carried several advanced instruments for studies of the ocean surface and atmosphere, operating at visible, infrared and microwave frequencies. The instrument complement included, amongst others, the Limb Infrared Monitor of the Stratosphere, LIMS, which was described earlier in this chapter, and the Stratospheric and Mesospheric Sounder, SAMS.

Another highly successful Earth observation mission was SEASAT (see Allan, 1983). Though this mission lasted only three months, it was able to demonstrate, in particular, the power of radar systems for all-weather monitoring of sea state and a wide variety of related phenomena. In its rather short life SEASAT was able to provide a copious quantity of data which demonstrated that the all-weather radar systems carried on board (including a Synthetic Aperture Radar, SAR, a Radar Altimeter and a Radar Scatterometer) were capable of delivering data which could be processed to provide valuable geophysical quantities. (The discussion of radar altimetry earlier in this chapter provides one example of the operation of such radar systems). Some of the scientific literature that has developed on the use of SEASAT data has shown SAR images of internal waves in shallow water, which reveal details of bottom topography; measurements of ocean surface topography by the Altimeter (see earlier); measurements of wave direction, wavelength and height from the Scatterometer and the Altimeter; and measurements of surface wind stress from the Scatterometer. SEASAT was a pioneering mission in that it was the first mission to clearly show the power of radar systems, particularly for operational applications where it is important not to be delayed in making observations by clouds. Many of the advantages do refer to near-real-time applications, which do not often arise in climate research; nevertheless, SEASAT showed also that the radars had a unique role to play in providing parameters such as surface wind stress, and ocean surface currents, which were not available using more conventional infrared or visible techniques. Thus, the real lesson of SEASAT as far as science is concerned is that we must utilize the full range of techniques available to us, which now includes radar methods, in order to gather as much information as possible from our orbiting observatories if we are to gain a full understanding of the climate system.

In the future, the next major missions in Earth observation will be the US Upper Atmosphere Research Satellite (UARS), scheduled at the time of writing to be launched by the Space Shuttle in the autumn of 1991, and the European ERS-1

Table 4.5 — Spacecraft characteristics

Spacecraft	Launcher	Size (m)	Weight (kg)	Payload and mass (kg)	Power[a] (W)	Data rate (Mb/s)	Altitude and orbit (km)	Scan type	Launch date
Geostationary									
Meteosat	Ariane	3.2 m × 2.1 m dia.	293	3-channel high resolution radiometer at visible, 6 & 11 μm. Mass 61 kg	280	0.166	36 000 (geostationary)	Spin scan at 100 rpm: 2500 (IR) or 5000 (Vis) lines image; 5 km or 2.5 km surface resolution	Various since Nov 1977
Polar orbiting									
Landsat 4/5	Delta	5.6 m × 2.2 m dia.	1938	Thematic mapper (TM), 7 bands from Vis to thermal IR (0.5–12 μm); Multi-Spectral Scanner (MSS), 4 bands from 0.5 to 1.1 μm. Mass 714 kg	1430	1506 (MSS) 85 (TM)	706 (98.2 deg. inclin.)	Cross-track scan, 185 km swath; 30–120 m surface resolution	July 1982 (Landsat 4)
SEASAT	Atlas	21 m × 1.5 m dia.	2290	5 sensors for ocean remote sensing (radar altimeter, scatterometer, synthetic aperture radar, microwave radiometer, and Vis–IR radiometer). Mass 714 kg	626	~100	800 (108 deg. inclin.)	Various; Nadir and Sidelooking radars, Visible, IR and microwave imagers and radiometers	June 1978
Nimbus-7	Delta	3.6 m × 1.6 m dia.	1021	Several instruments for remote sensing of the stratosphere: —Coastal Zone Colour Scanner (CZCS) —Earth Radiation Budget (ERB) —Limb Infrared Monitor of the Stratosphere (LIMS) —Stratospheric and Mesospheric Sounder (SAMS) —Stratospheric Aerosol Measurement (SAM II) —Solar Backscatter Ultraviolet Spectrometer (SBUV) —Scanning Multifrequency Microwave Radiometer (SMMR) —Temperature Humidity, Infrared Radiometer (THIR). Mass 406 kg	310	0.8	950 (99.3 deg. inclin.)	Nadir and Limb sounding	Oct 1978
NOAA/TIROS	Atlas	4.18 m × 1.88 m dia.	2265	Mass 303 kg —TIROS Operational Vertical Sounder (TOVS)[b] —Advanced Very High Resolution Radiometer (AVHRR) —Earth Radiation Budget Experiment (ERBE) —Solar Backscatter Ultraviolet Experimental (SBUV) —Search and Rescue —Data Collection and Relay. Mass 475 kg	525	2.66	870 (98.6 deg. inclin., sun-synchronous)	Cross-track, swath width: 2800 km (imager); 2347 km (sounder); 1, 4, 17, 109 and 147 km surface resolution, depending on sensor	TIROS-N pre-operational mission, Oct 1978; NOAA-A, June 1979; NOAA-G, Sept 1986 (data quoted here for NOAA-H)
UARS	STS	9.8 m × 4.3 m dia.	6736	9 instruments for study of the stratosphere (see Table 4.6). Mass 2283 kg	1600	32	600 (57 deg. inclin.)	3-axis, stabilized by reaction wheel and magnetic torquers (0.01 deg. precision)	1991
ERS-1	Ariane	3 m × 2 m dia.	2283	3 core radar instruments for ocean surface studies (altimeter, SAR and Scatt) plus ATSR IR & MW sounders and ranging experiment. (Table 4.7) Mass 950 kg	2000	100	777 (99 deg. inclin.)	3-axis stabilised; various scan types	1991
Polar platform	STS	12 m × 4.3 m dia.	12210	'Observatory' with large number of instruments for remote sensing of the climate system (see Table 6.1). Mass 3000 kg	6000	50	705 (sun-sync.)	3-axis stabilised; various scan types	Series beginning 1997: EOS A, EOS B, ESA, PPF

[a] At the Beginning of Life (BoL).
[b] TOVS comprises:
—High Resolution Infrared Sounder (HIRS) — 20 channels (0.7–15 μm); 17 km surface resolution; temperature, humidity, ozone
—Stratosphere Sounding Unit (SSU) —3 channels (15 μm); 147 km surface resolution; stratospheric temperature
—Microwave Sounding Unit (MSU) — 4 channels (50–57 GHz); 109 km surface resolution; tropospheric temperature in cloudy areas.

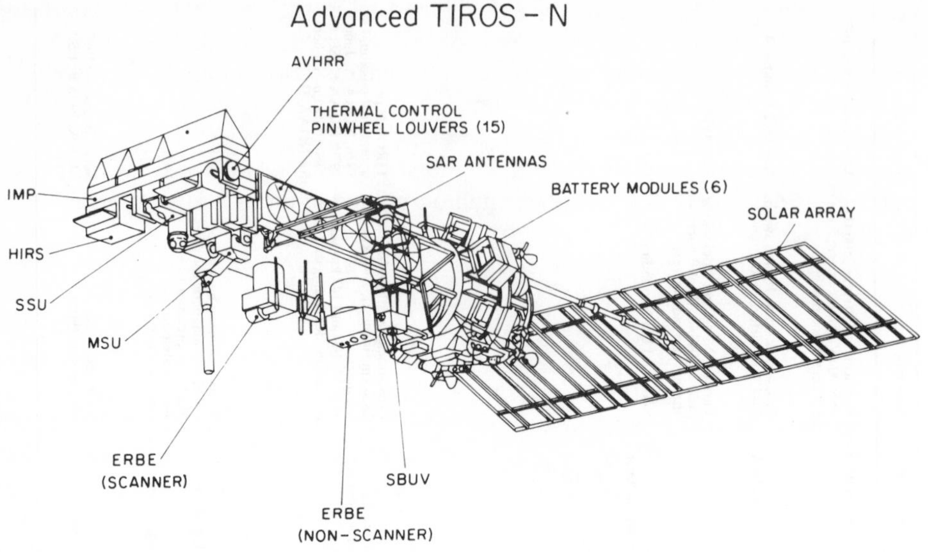

Advanced TIROS – N

Fig. 4.16 — The TIROS-N spacecraft.

mission , to be launched by Ariane also in 1991. Fig. 4.18 shows the scale of the UARS spacecraft compared with other spacecraft systems. Strictly speaking, the actual UARS 'bus' spacecraft comprises only the rectangular components at one end of the spacecraft: the remainder is entirely made up of payload instruments and the structure required to support them. UARS will carry a payload of 9 different scientific experiments, aimed at studies of the dynamics, chemistry and energy balance of the stratosphere and mesosphere.

The list of instruments to be flown on UARS is shown in Table 4.6, along with their primary goals. It must be stressed that these instruments represent very advanced technology and sophisticated designs, and in many cases are quite new instrument concepts that have been designed and developed specifically for the UARS mission. This illustrates two points: first that we are some way from having reached the ultimate design and performance of remote sensing instruments, which have considerable development potential still; and, second, that the UARS programme, including instruments of the most advanced design which have never before been flown in space , is an extremely ambitious programme. However, in order to extract the full potential of global remote sensing from space, we must continue to be far-sighted in our ambitions. Many challenges of technology and of sheer human ingenuity face us if we are to achieve that full potential.

One of the instruments for UARS, the Improved Stratospheric and Mesospheric Sounder (ISAMS for short!) is a very advanced design of sensor (Taylor, 1987). It comprises some 12 or so separate spectral 'channels' each of which is tuned to sense infrared signatures from the atmosphere from a specific gas in the atmosphere. This is achieved by a technique called pressure modulation radiometry (cf. Fig. 4.5), in which a sample of the gas itself is actually carried in the instrument on board the satellite; this is used to filter out from all the different wavelengths of radiation emitted by the atmosphere only those wavelengths corresponding to the gas in question. By this means the intensity of those spectral signatures can be measured by sensitive infrared detectors in the instrument, and the amount of the gas in the atmosphere calculated. The technique is so sensitive that trace gases with relative concentrations as low as 1 part in one million-million of the pressure at sea level can be detected.

Most of the other experiments on UARS are being provided by American groups through extensive support provided to academia and industry in the USA by NASA; three of them, the Microwave Limb Sounder (MLS), the Halogen Occultation Experiment (HALOE), and the High Resolution Doppler Imager (HRDI) involve British groups, the first through the provision of some very high technology components, and the other two through involvement of individual British scientists in their supporting science teams. Fig. 4.19 shows a schematic diagram of the UARS spacecraft, identifying the position of the various sensors, including those just mentioned. The MMS (Multi-Mission Spacecraft) is the supporting bus, shown at one end of the spacecraft. Plate 3 shows an artist's impression of UARS in orbit. At the time of writing (spring 1990), the UARS instruments are being integrated onto the spacecraft by the General Electric Corporation in the US, on behalf of NASA.

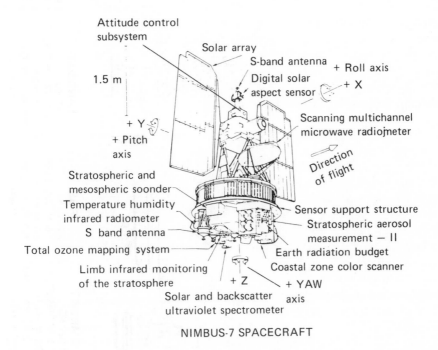

Fig. 4.17 — The Nimbus-7 spacecraft.

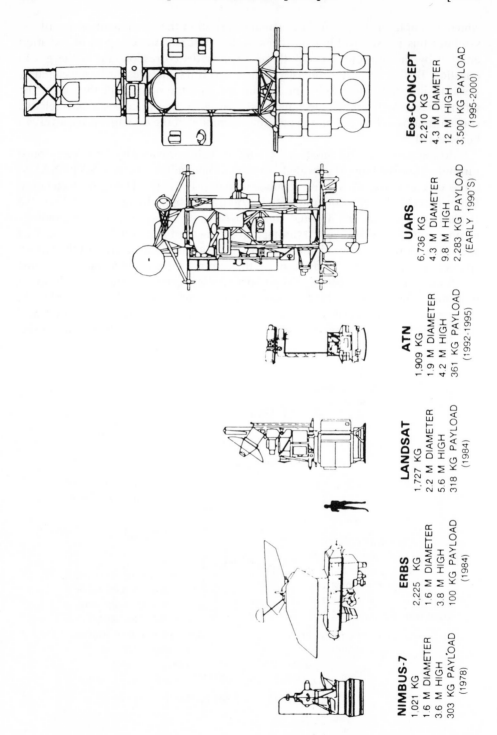

Fig. 4.18 — The development of spacecraft up to the polar platform of the Space Station, shown at right.

Table 4.6 — The UARS payload: Summary of UARS instrument investigations

Instrument	Description	Measurement objectives	Remarks	Principal investigator, Institution
CLAES: Cryogenic Limb Array Etalon Spectrometer	Scanning spectrometer sensing atmospheric infrared emissions in the spectral range 3.5–12.7 microns	Concentrations of members of the N and Cl families, O_3, H_2O, CH_4, and CO_2 at altitudes of 10–60 km; atmospheric temperature profiles for indirect wind measurements	Detectors cooled by solid Ne at −260°C, optics cooled by solid CO_2 at −150°C	A. E. Roche, Lockheed Palo Alto Research Laboratory, USA
ISAMS: Improved Stratospheric and Mesospheric Sounder	Radiometer sensing atmospheric infrared emissions in the spectral range 4.6–16.6 microns	Atmospheric temperature structure and variability; minor constituent distributions including the N family, water, methane, carbon monoxide, and ozone.	Pressure modulated gas filters; detectors cooled to −195°C by closed-cycle Stirling refrigerator	F. W. Taylor, Oxford University, UK
MLS: Microwave Limb Sounder	Radiometer sensing atmospheric microwave emissions at frequencies of 63, 183, and 205 GHz.	Concentrations of ClO, H_2O, O_3, and atmospheric pressure at various altitudes from 5 to 85 km	Will furnish first global data set on ClO, a key reactant in catalytic O_3 destruction	J. W. Waters, Jet Propulsion Laboratory, USA
HALOE: Halogen Occultation Experiment	Radiometer sensing atmospheric infrared absorptions from occulted sunlight in the spectral range 2.43–10.25 microns	Vertical distributions of HCl, HF, CH_4, CO_2, O_3, H_2O, and members of the N family	Utilizes gas filters, makes 28 solar-occultation measurements per day	J. M. Russell, NASA Langley Research Center, USA
HRDI: High Resolution Doppler Imager	Fabry–Perot interferometer sensing atmospheric emission and absorption in visible and near-infrared spectral ranges	Velocity of upper-atmosphere wind field through measurement of Doppler shifts of molecular absorption lines (below 45 km, daytime only) and atomic emission lines (above 60 km, day/night)	Yields direct measurements of wind field; initial flight demonstration of Doppler measurement technique in the stratosphere	P. B. Hays, University of Michigan, USA
WINDII: Wind Imaging Interferometer	Michelson interferometer sensing atmospheric emissions in visible and near-infrared spectral ranges	Velocity of upper-atmosphere wind field through measurement of Doppler shifts of molecular and atomic emission lines above 80 km	Yields direct measurements of wind field	G. G. Shepherd, York University, Canada
SUSIM: Solar Ultraviolet Spectral Irradiance Monitor	Full-disk solar ultraviolet irradiance spectrometer	Spectrum of solar ultraviolet radiation from 120 to 400 nm, with resolution of 0.1 nm	Incorporates four deuterium calibration lamps to verify long-term instrument and detector stability	G. E. Brueckner, Naval Research Laboratory, USA
SOLSTICE: Solar/Stellar Irradiance Comparison Experiment	Full-disk solar ultraviolet irradiance spectrometer	Spectrum of solar ultraviolet radiation from 115 to 430 nm, with resolution of 0.12 nm	Compares solar ultraviolet output with ultraviolet radiation from bright blue stars for calibration	G. J. Rottman, University of Colorado, USA
PEM: Particle Environment Monitor	Electron, proton, and imaging X-ray spectrometer	Energy spectrum of electrons (1 eV–5 MeV), protons (1 eV–150 MeV) and X-rays (2–50 keV)	Particle measurements made in vicinity of spacecraft; will also determine energy deposition by high-energy electrons.	J. D. Winningham, Southwest Research Institute, USA

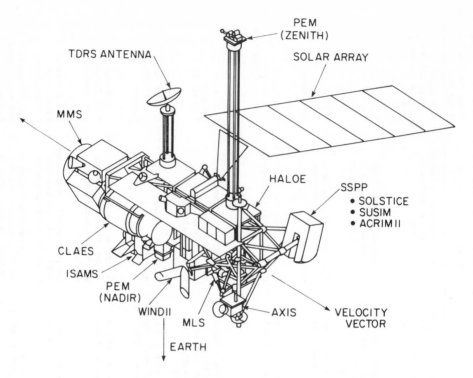

Fig. 4.19 — The Upper Atmosphere Research Satellite (UARS).

The second major forthcoming Earth observation mission which we mentioned above represents an important development in European interest in the study of our planet and its climate from space. This is the first European remote sensing satellite, known as ERS-1. ERS-1 is a mission devoted to the study of the oceans and to a lesser extent the land surface, and their role in the global climate. ERS-1 is also intended to provide a stimulus to European industry in the drive to use remotely sensed data from space in the development of commmercial and practical applications, such as land use planning, fisheries, ocean routing, and so on. It is the scientific application to the global climate problem which is most of interest to us in this book, of course, and which we will concentrate on. The spacecraft 'bus' for ERS-1 is the French SPOT satellite (Système Pour Observation de la Terre). ERS-1 will carry three radar systems (though some systems are shared between two of the radars), the Synthetic Aperture Radar (SAR), the Scatterometer (SCATT) and the Radar Altimeter (RA). These instruments will permit a wide variety of ocean surface parameters to be measured, in all weathers. These parameters include, as we have seen above, surface winds, ocean currents, and wave characteristics (wave speed, direction, frequency spectrum, etc.). The RA will also give information on the global gravity field, from the effect that variations in the gravitational potential has on the ocean surface topography around the orbit.

In addition to these 'core' instruments, there will be two nationally funded instruments flying on ERS-1, namely a Germany-provided radio-ranging experiment to improve the accuracy with which the spacecraft position is known, and an experiment to measure the temperature of the ocean surface with high accuracy. This latter experiment is being provided by the UK, and is called the Along-Track Scanning Radiometer, and works in the infrared region of the spectrum to sense thermal radiation emitted by the ocean (see Edwards *et al.*, 1990; Harries *et al.*, 1983). A number of novel technology features, and developments in the scientific interpretation of the data, mean that the temperature of the surface of the oceans will be measured globally to an accuracy of a few tenths of a degree Celsius: to do that from an orbiting spacecraft, nearly 1000 km away in space is a formidable challenge, but it now looks as though that challenge has been overcome. A full list of the ERS-1 instrumentation and its capabilities is given in Table 4.7: it is planned that this will be the first European satellite for studies of the Earth in a series, continuing in 1994/95 with a repeat version, ERS-2, and in the late 1990s with even more advanced instruments on the European and other elements of the Space Station.

Plate 4 shows a picture of the configuration of the ERS-1 satellite. The three radar systems, the SAR, scatterometer, and the altimeter can be seen towards the top of the spacecraft, while the Along-Track Scanning Radiometer and its associated Microwave Sounder (provided by French groups) can be seen 'inboard' of the SAR antenna. The satellite will be oriented with the altimeter/ATSR face pointing down towards the Earth, and with the velocity vector along the long axis of the SAR antenna.

We have referred to the class of geostationary satellites already, and the European contribution to the set of geostationary meteorological satellite that circle the Earth, hovering above points on the equator. Meteosat is a spin stabilized spacecraft, carrying a three-channel radiometer which builds up a picture of the Earth in a 'spin-scan' technique, in which one line of the image is built up every time the satellite rotates. An east-west line of picture elements is built up as the spacecraft rotates, and a succession of lines is generated by moving a scan mirror in steps from south to north during the non-Earth pointing part of the scan. The three spectral channels include one in the visible, and two in the infrared, in the bands 5.7–7.1 microns and 10.5–12.5 microns. The band centred at 6 microns is sensitive to water vapour in the mid-troposphere, while the longer wave channel is sensitive to surface radiation, attenuated slightly by the troposphere. Fig. 4.20 shows a drawing of the Meteosat satellite: it is cylindrical, with a diameter of 2.1 m, and rotates in orbit about the axis of the central antenna. The three-channel radiometer is contained within the body of the spacecraft, and views the atmosphere once per rotation via the aperture shown. The exterior of the body of the satellite is covered in solar cells to provide power. The photograph shown in Plate 1, illustrates the images obtainable from Meteosat in the global, visible mode: Plate 5 shows a close-up of the European region taken from Plate 1, so familiar now from our TV weather bulletins; and Plate 6 illustrates some of the data which is obtained using the 6 microns 'water vapour' channel: this shows regions of high and low humidity in the mid-troposphere, and may be used to track air motions at these altitudes. This is what the Earth would look like to a being whose eyes had developed so as to be tuned to wavelengths about a factor of ten longer than our own.

Further into the future, of course, the next major Earth observing space system

Table 4.7 — The ERS-1 payload

	Instrument	Major modes	Wavelength/frequency	Objective
Core payload (ESA provided)	AMI	Synthetic Aperture Radar	5 GHz (C band)	High-resolution radar images
	AMI	Wave Mode Wind scatterometer		Directional wave spectra Oceanic wind fields
	RA	Ocean Mode	13.6 GHz (K_u band)	High-precision altitude Echo characteristics
		Ice Mode		Radar echoes over rough surfaces
	Laser retroreflector		Visible	Target for laser tracking stations
Announcement of opportunity payload (nationally provided)	ATSR	Along Track Scanning Radiometer	3.7, 11 & 12 μm	Optical/IR measurements of sea-surface temperature
		Microwave radiometer	23, 35 GHz	Atmospheric water vapour measurements
	PRARE		RF	High-precision measurements of satellite range and range-rate from special ground stations

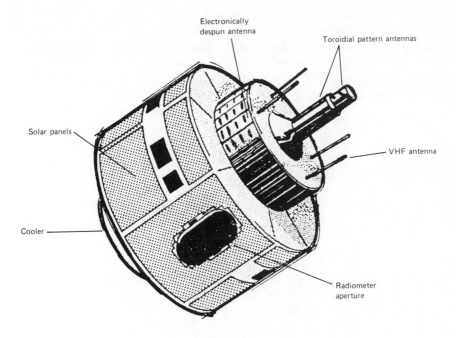

Fig. 4.20 — The Meteosat spacecraft.

will be the Space Station, now called 'Freedom' by NASA. Space Station Freedom will comprise a main component in near equatorial LEO, which will include (at least on present planning) a large framework supporting a number of manned modules, large solar arrays, various instruments and other vehicles. There will also be a polar orbiting element which represents the most important component as far as Earth observation interests are concerned, namely the so-called polar platform, the design of which is still being studied by space agencies in the USA, Europe and Japan. The baseline design parameters being used by the space agencies are noted in Table 4.5, though it must be stressed that these are still somewhat provisional, and depend on adequate funding as well as the necessary technical progress being made. The capabilities of the polar platforms in terms of payload mass carrying ability, data transmission rate and so on are enormous, but are demanded by present-day modern remote sensing equipment. The concern which many of us share these days is not that we cannot design the system to properly monitor our planet from space, but that we are not giving adequate priority to the training of more scientists to be able to make sense of the masses of new data that will come flooding back from our new sensors, flying on the various Space Station elements, especially the polar platforms.

There are a wide variety of techniques which have been developed by scientists around the world over the past 20 years or so, each adapted to particular needs for observations of various parameters with assorted space and time scale requirements. We do not claim here a comprehensive survey, which would itself take up a whole volume, and with regret we will exclude any discussion of such aspects as active laser probing, of synthetic aperture radar methods, and several others. What we have tried to do is to give sufficient insight to the reader into some of the techniques which are of particular relevance to studies of the climate system, in order to understand some of the principles and to encourage further reading. For lidar methods see, for example, Russell *et al.* (1979); for synthetic aperture radar methods see Ulaby *et al.* (1981, 1982 and 1986).

5

Space observations — some examples

In this book so far, we have surveyed the elements of the theory of climate; we have discussed some specific popular climate issues, such as the ozone hole, the greenhouse effect, and so on; and in the last chapter we reviewed some of the principles of remote sensing from space which particularly relate to climate studies. In this chapter we will try to bring these separate themes together, by illustrating how some of these problems have been addressed through the use of satellite-based Earth observing techniques. These examples are drawn from a variety of sources, but we must particularly pay tribute to the major role played by NASA in the United States in the development and realization of this subject: the considerable investment that this agency has made in the development of space techniques for the study of the well-being of our planet has provided spacecraft and launchers, instruments and technology, and computers and scientists to ensure the adequate exploitation of the data obtained. Notwithstanding the great wealth of the USA which makes this investment possible, it has required considerable wisdom and insight to ensure that this wealth is invested in this way: NASA expenditure in Earth observation from space has indeed been an investment in our future.

Thus, many of the examples that the reader will encounter in the following pages have been drawn from NASA programmes. However, it should not be imagined that other countries have played no role at all: indeed, this is far from the truth. Several European countries, and perhaps most notably the UK, France and Germany, have contributed to the development of new techniques and methods of remote sounding which have often been very ingenious in their conception. Flight instruments have been developed under national funding and flown on American spacecraft, and underpinning programmes have been undertaken on instrument R&D, aircraft and balloon test flights, laboratory studies, as well as the background geophysics. Other nations, such as the USSR, Japan, India and China, amongst others, have mounted their own programmes in Earth observation, though they have tended to follow somewhat behind in the development of advanced sensors (until now: several countries, for example Japan, show signs of increasing interest in the development of their own advanced instrument programmes). The European Space Agency, while perhaps arriving a little late on the scene, is now getting involved in Earth observation in a major way. The ERS-1 satellite, as we saw in Chapter 4, represents an important new development in Earth observation, especially in the use of radar techniques, and in the flight of advanced sensors like the ATSR infrared radiometer. Present plans call for an extension of these activities to ERS-2 and the polar platform, and the Agency looks set to become an important authority on satellite observations of the Earth in the future.

Let us, then, take a look at the data products which have been derived from current satellite systems, and try to relate what the sensors tell us about the scientific processes that are at play in the climate system, drawing on what we have learned in Chapters 2, 3 and 4. Where relevant, we will try also to describe forthcoming projects, and where they might lead us.

5.1 SATELLITE OBSERVATIONS OF OZONE AND OTHER TRACE CONSTITUENTS IN THE STRATOSPHERE

5.1.1 Ozone hole observations

Chapter 3 described our current scientific understanding of how the annual depletion of ozone over the Antarctic arises (and probably also over the Arctic, though to a lesser degree), and told the story of how the discovery of the Antarctic hole occurred. As described there, this was not initially a success story for space observations, since the satellite observations missed the onset of the phenomenon, while more humble ground-based techniques scored a major triumph. But, as we saw in Chapter 3, this was not so much due to the inadequacy of satellite techniques, but the mis-application of such techniques by human beings. The satellite system operators decided to instruct the analysis system to ignore data which showed 'anomalously' low levels of ozone, and indeed, who could blame these people, when nothing like the 'hole' phenomenon had been seen previously? Once the mistake had been recognized, and once the satellite data were freed from the artificial constraint in the processing software, the space system became a powerful and precise technique for tracking the phenomenon, and for providing information critical to its understanding, as we shall see.

As of the mid-1980s, and indeed still now, the only satellite-based techniques for monitoring ozone in the stratosphere which was in regular use, and could be in any way described as an 'operational' system, were the Total Ozone Monitoring Spectrometer (TOMS), and the Solar Backscatter Ultra Violet spectrometer (SBUV), flown on Nimbus-7 and subsequently on a series of NOAA weather satellites. The instruments detect the backscatter of solar radiation from the atmosphere, which at short wavelengths depends on the amount of ozone present. The method operates only during daylight, and does not give data over the poles during the periods of total darkness (i.e. in the depth of mid-winter). Nevertheless, the availability of daylight, and the consequent coverage of TOMS are sufficient to observe the Antarctic depletion (which, the reader will recall, does not commence, anyway, until sunlight returns to illuminate the south polar vortex region (see Chapter 3)). An alternative technique, used in the Stratospheric Aerosol and Gas Experiments (SAGE) I and II, employs the solar occultation (limb sounding) technique (see, for example, McCormick *et al.*, 1989). This technique has also yielded valuable data to high southern latitudes for the periods 1979–1981 and 1984 to at least 1989.

Before the discovery of the ozone hole, the main interest in satellite ozone measurements was to attempt to detect any long-term changes in the concentration of ozone, worldwide, which could be due to the chemical imbalancing of the stratospheric photochemical 'soup' through pollution by chlorofluorocarbons (CFCs) and the resulting free chlorine, or nitrogen oxides, or any other photochemically active pollutant. Trends of the order of a few per cent per year were being sought, and indeed this remains an important research goal, though rather overshadowed by the polar phenomenon, which is, of course, far more dramatic than this, with up to 90 per cent of the ozone column disappearing on a local basis within a time scale of about one month! This primary goal of detecting trends in ozone does rather vindicate the operators who were caught slightly 'with their trousers down' when the ozone hole hit us.

An example of the type of data obtained from the TOMS experiment was given earlier in Fig. 3.3 in Chapter 3, which shows one of the deepest depletions of ozone seen, on the 5 October 1987. The whole of the Antarctic continent is covered by a vast 'hole' in the ozone, with less than 200 Dobson units (DU) observed over the whole continent, and minimum ozone levels of around 125 DU showing over the pole. This is to be compared with a band of high ozone immediately surrounding the depletion area of up to 450 DU, and with levels further north, and nearer the equator, of 250 DU or so.

The power of satellite data can be seen in the next illustration, Plate 7, which shows a fully developed Antarctic ozone hole in 1983 from Nimbus-7 TOMS data. The development of the full two-dimensional dynamics of the process can be mapped out by these data, and the relationship between the low ozone amounts and these dynamical effects can be studied. It should be remembered that it is the development of a well-defined vortex of cold, dark air that is the precursor of the catastrophic ozone loss when the Sun returns to the southern pole in the spring, and so the dynamical motions that can be tracked by such data are an essential element of the total description of the phenomenon, as we saw in Chapter 3.

The existence of the high ozone 'collar' around the depletion zone is of interest, since it appears to be a well-established phenomenon from the evidence of the satellite data. Plate 8, for example, shows a series of October averages for the years from 1979 to 1986, in which the collar can be seen as a repeatable and persistent observation. This diagram also illustrates that the growth of the deep hole is a very recent event, coinciding closely in time with the known growth of chlorofluorocarbons in the atmosphere.

Scientists have also been interested to establish whether the annual repetition of the ozone hole phenomenon is causing any long-term accumulative fall in global ozone levels, or whether the continuous cycle of ozone formation and destruction 'repairs' all the damage caused by the springtime depletions. Several studies (see Chapter 3 — Proffitt et al., 1989; Atkinson et al., 1989; NASA, 1988c) suggest that a gradual depletion of ozone at other latitudes is taking place, though strictly, of course, we have no way of knowing if this is due to the ozone hole phenomenon, or to other processes including background natural fluctuations. Recent evidence (Proffitt et al., 1989) indicates the possibility that ozone destruction occurs just outside the polar vortex, during the (southern) winter preceding each ozone hole event. Such an effect should be visible from global satellite data if it exists.

The incentive to continue TOMS observations from future operational satellite systems, whether dedicated free-flying spacecraft or as part of the polar platform complex, is clearly very great. In addition, the Upper Atmosphere Research Satellite to be launched in 1991 (and described in Chapter 4), will carry a number of highly specialized instruments for measuring ozone, related chemicals, dynamical motions, aerosols and so on, which will permit a very detailed and highly focussed study of the complete cycle of processes that make up the ozone hole phenomenon. Both operational 'monitoring' observations and these more focussed studies of the complete global ozone question, not only the polar manifestations, will be important elements of our future use of space systems to protect our planet.

5.1.2   Global ozone and other trace constituents measured by LIMS
Nimbus-7 also carried other instruments for measuring stratospheric ozone and other chemically and radiatively important trace constituents. One of them was the LIMS experiment, described in Chapter 4. Though this instrument operated (as planned) for only 7 months, after which time its on-board cryogen ran out, it has provided a very valuable data set on a number of important chemical species. LIMS has particularly thrown light on the global distribution of the concentrations of these species as a function of latitude, and their variations with time.

To illustrate these data we can show Plates 9, 10 and 11, kindly provided by Dr J. M. Russell of NASA's Langley Research Center in the USA (see Russell *et al.*, 1984a; Gille *et al.*, 1984; and Remsberg *et al.*, 1984). The first of these figures gives the zonal mean (i.e. averaged over all longitudes for a given latitude band) cross-section of the measured ozone mixing ratio for the months of January and May, 1979. The units are parts per million by volume, ppmv. More dramatic variations with time can be seen in the measurements of nitric acid, HNO_3, for the same months, as shown in the second figure, Plate 10, in which the units are 1000 times smaller, parts per billion by volume. Here it can be seen that the high winter pole concentrations reverse completely as the seasons reverse. As a third example, Plate 11 shows similar data for water vapour, H_2O, this time in the same units as ozone. The strong drying influence of air rising into the tropical stratosphere through the high, cold tropopause at low latitudes can clearly be seen, and also the gradual increase in mixing ratio with height in the stratosphere, due probably to the formation of water molecules by the oxidation of methane, CH_4, reacting with atomic oxygen. The figure also shows interesting 'sources' of H_2O at high southern latitudes in May, and at high altitudes in both January and May.

It is clear from these three examples that global satellite data give us a very powerful insight into the global distributions and variability of trace chemicals, which can be used to make very stringent tests of the predictions of theory by comparison of the measurements with model results. Many further examples of the use of LIMS data for geophysical and climate studies exist int the literature (e.g. Russell *et al.*, 1984b; Solomon *et al.*, 1986; Harries, 1982b; Pyle *et al.*, 1983; Jones R. L. *et al.*, 1986).

5.2   SATELLITE OBSERVATIONS OF SEA SURFACE TEMPERATURE
The temperature of the Pacific Ocean surface was shown in Chapter 3 to be a direct signal of 'El Nino–Southern Oscillation' events, which we now know can have major

implications for the global climate. We also know that the ocean surface can act in two ways to modulate and moderate the global greenhouse effect, either by absorbing greenhouse gases from the atmosphere or by direct thermal exchange with the atmosphere. More generally, the thermal mass of the surface layer of the ocean, and particularly the topmost few tens to hundreds of metres, represents a major element in the global energy balance of the planet, and the temperature of this layer compared with that of the overlying atmosphere helps to determine the amount of latent heat energy released during evaporation processes. Each of these aspects has been considered in earlier chapters, and each demonstrates that the sea surface temperature (SST) is a very important parameter to monitor globally.

The technique that is used, of course, to measure SST is to observe the intensity of thermal radiation emitted by the ocean surface in the infrared or the microwave regions of the spectrum, after making due allowance for the absorbing properties of the atmosphere in between the observer and the surface (see Chapter 2). Several different specific sensors have been flown on spacecraft to date, not all of which we can report on here, but as an example, SST data are operationally available from the Advanced Very High Resolution infrared Radiometer (AVHRR), which is flown as part of the meteorological sounding package on the NOAA satellite series (Chapter 4). The infrared systems, as epitomized by AVHRR, usually make measurements at more than one wavelength, and use the differential absorption properties at the different (usually two) wavelengths to eliminate the effects of the atmosphere, at least to first order (Barton and Cechet, 1989). The microwave techniques usually directly measure the intensity at a few wavelengths and employ a direct data fit, not relying as much as in the IR on differential absorption properties. The infrared enjoys the advantage that the emissivity of the ocean is very close to 1.0 , so that minimal confusion occurs from reflected signal, but it suffers from not being cloud-transparent. The microwave region has the reverse attributes: the surface emissivity is nearer 0.5, while the presence of all but the heaviest rain clouds is of little concern.

Most sensors to date have been designed particularily to achieve high relative accuracy in temperaure measurements, in order to be able to recognize and monitor temperature gradients in the horizontal, for example in the form of oceanic fronts. This has in practice been rather an enforced choice, since the problems of making measurements with high absolute accuracy are considerable. The climate problem, however, requires absolute accuracy more so than relative accuracy, and does not require a terribly high spatial resolution, since we are concerned with a global problem, and small local variations are, to a good approximation, to be discounted as 'internal fluctuations'. Thus there is a trade-off possible between high spatial resolution and low absolute accuracy on the one hand , and low spatial resolution and high absolute accuracy on the other, and this has been the design philosophy behind the Along-Track Scanning Radiometer (ATSR), which was described in Chapter 4, and which will be flown on the first European remote sensing satellite, ERS-1. The ATSR employs multi-wavelength and multi-angle correction techniques for removing the effects of the atmosphere, cold detectors for high sensitivity, and averages over a fairly large (50×50 km^2) area, all to achieve the ultimate in absolute accuracy from a spaceborne infrared sounder.

However, the AVHRR instrument has been the main source of virtually continuous SST data from space over the past decade or so. Plate 12 shows some of the data from this instrument which illustrates the high relative accuracy of the device in detecting horizontal gradients of temperature. It also shows what a problem the presence of clouds in the field of view can pose: for instance, the difficulty of identifying the boundaries between sea and low-lying cloud with very similar temperatures can be considerable. This problem is particularly acute over ice and snow surfaces, where the temperaure and emissivity properties can be so similar that discrimination between cloud and surface must be done by using independent techniques.

Some work has been done in several centres worldwide to evaluate data on a more global basis, as required by global climate researchers. Perhaps the best known example of such work to date is that by Chahine and Susskind in the USA (e.g. see Chahine *et al.*, 1986), which has taken global data sets from two sensors, one an infrared device, the High Resolution Infrared Sounder (HIRS), and the other the Microwave Sounding Unit (MSU). Though the MSU provides somewhat lower accuracy than the HIRS device, particularily at the boundaries of oceanic regions where the very broad beam pattern (diffraction side-lobes) causes serious interference from the adjacent land surfaces, nevertheless, the MSU permits HIRS data gaps due to clouds to be filled in. In this work, a careful synthesis of the two data sets has been carried out, and an example of the resulting combined data set is shown in Plate 13, representing the global distribution of SST for a completely cloud-free globe. While the absolute accuracy is not high, especially over land where surface emissivity effects are important, the resultant data set clearly shows the potential of satellite data: to test global climate models we require data sets like this one, with high absolute accuracy and good time repetition (say weekly).

In the future, we have already discussed in several places the role of the ATSR experiment on ERS-2 in improving absolute accuracies of SST globally. The operational data from AVHRR will continue for the foreseeable future, and plans include a programme of improvement to the basic sensor. In the polar platform/Space Station era, there is an intention to fly several sensor systems which will measure, amongst other things, SST, some at high spatial resolution (a few $\times$ 100 m), others with medium spatial resolution (a few km), both in the infrared and the microwave regions. Further details on possible instrumentation for the polar platforms of the Space Station are given in Chapter 6 (see Tables 6.1, 6.2, 6.3).

5.3 SATELLITE OBSERVATIONS OF POLAR ICE

The 'cryosphere' is an important component of the climate system, as we discussed in Chapter 2. The polar ice sheets, especially in the south, contain a vast quantity of water which would otherwise be available to produce much higher sea levels. The massive volume of ice also represents a very large thermal 'ballast', in that a large supply of heat energy could be absorbed by (or extracted from) this ice if the climate were to fluctuate; the existence of the ice represents a stabilizing influence on the global climate system. Furthermore, we saw when describing the Antarctic ozone hole that the presence of a high and extremely cold polar continent gives rise to a strong circumpolar vortex, which in turn gives rise to conditions under which chlorine can destroy ozone particularly effectively. So, the cold, icy regions of the world are of direct importance to the global climate system in a number of ways.

The principal components of the cryosphere were identified in Chapter 2, see for example Table 2.1, and comprise the great Antarctic and Greenland ice sheets, the extensive sea ice cover of the Arctic Ocean, together with that surrounding the Antarctic continent, plus much smaller global contributions to the ice/snow inventory coming from glaciers, and seasonal snow and ice. There are two major difficulties associated with the remote sensing of the cryosphere and its components from space. These are, first, the frequent presence of cloud, often persistent in nature, which makes the use of infrared or visible observations ineffective, and which therefore takes us in the direction of utilizing microwavelengths in order to make observations through the cloud cover. Second, if we do use infrared or visible techniques, there is often great difficulty in discriminating between cloud, which is, of course white and cold, and the surface, which is also white and cold. In some applications, for example the study of the net radiative energy balance, we may not be concerned with such discrimination, being preoccupied only with the net radiative fluxes, and not the detailed components of the flux. However, there are obviously other cases in which it is important to be able to discriminate between surface and cloud effects. This second problem also argues for the use of cloud-penetrating microwave sounding methods from space.

Cryospheric remote sensing is, therefore, the domain in which the development of microwave techniques has perhaps enjoyed its greatest stimulus, and so it is not surprising that the examples of the study of the cryosphere from space that we now come to are drawn from such techniques. It has been demonstrated that both passive, radiometric techniques, sensitive to surface temperature, ice condition, moisture, etc.; and active radar methods, both imaging and non-imaging, can be applied to cryospheric climate problems, with great success. We will now take a look at results from both types of observations.

An instrument called the Electrically Scanning Microwave Radiometer was flown on the Nimbus-5 spacecraft by NASA in the mid-1970s. Data from this instrument have been analysed by a number of workers, for example Meier (1983), and an example of the results obtained is shown in Plate 14 (see also Fig. 2.27). This records the extent (not the thickness or volume, of course) of ice and snow as detected by the microwave radiometer on board the spacecraft, and clearly shows the seasonal variation in the southern hemisphere. Since the variations at the perimeters of both north and south polar regions occur principally in sea ice, rather than thick ice sheets, and since the sea ice is thin (typically of the order of metres) the measurement of sea ice extent, seasonal variability and any longer-term trends provides a very sensitive test of climatic change. For the same reason, sea ice extent is a sensitive parameter for testing climate models: thus, it can be expected that responses to simulated perturbations of the climate will show up in model runs quite effectively. For these reasons, the sort of data shown in Plate 14 are of great value to climate research, and further developments in the remote sensing techniques and technology are deserving of considerable further investment by funding agencies.

The use of radar techniques, in which a powerful microwave pulse is reflected from the surface of the ice or snow under study, rather than relying on the sensing of the natural thermal radiation from the surface itself, provides a complementary tool to the passive methods. Radar techniques are not so sensitive to surface physical properties, such as temperature, for example, but are sensitive to mechanical surface properties, such as roughness. They also allow the use of time-of-flight methods to determine distances to great accuracy, and can therefore be used to measure surface topography across the great ice sheets. It is from this latter class of observations that we have taken our example here.

The radar altimeter is a device which focusses on the measurement of distance, and so uses a very short pulse of radiation, typically at frequencies between 3 and 30 GHz. For example the radar altimeter on the SEASAT spacecraft (see Chapter 4 for more details of SEASAT) operated at a frequency of 13.5 GHz (corresponding to a radar wavelength of about 2 cm). By the use of sophisticated electronic processing, and through the application of careful corrections for the effect of the atmosphere on the radar pulse transit time, it was possible to obtain a measure of the absolute surface topography below the spacecraft to an accuracy of a few metres; the relative accuracy of the altimeter, i.e. the measurement of relative position, which could then be used to determine relative topography along track under the satellite, was an incredible 10 cm or so. At a range of close to 1000 km, this was a remarkable achievement.

Fig. 5.1 shows some processed data from SEASAT. These are absolute surface topographies for Greenland and part of Antarctica (complete coverage to the pole was limited by the inclination of the orbit). The sheer power of space techniques, which are capable of producing data sets like these, which would take ages by conventional surface surveying procedures, is convincingly demonstrated by these results. Measurements of ice and snow topographies can be used to provide quantitative data for models of ice and snow amounts, distributions and seasonal variations, which are essential for the accurate simulation, and ultimately prediction of climate trends. Also, the data can be used in a more direct way to monitor variations, both seasonal and longer-term trends, to give warning signals of any actual climatic changes. Since the topography of the ice sheets does not change rapidly, it is sufficient to obtain a complete global 'inventory' of the type shown in Fig. 5.1 perhaps once every few years or so.

5.4 SATELLITE OBSERVATIONS OF THE OCEAN SURFACE

We saw in Chapters 2 and 3 that the impact of the oceans on the climate system is considerable, and that this impact manifests itself through interactions between the ocean surface and the atmosphere. The oceans respond to the wind stress, as it is called, as winds blow over the oceans, and drive water before them, in some cases inducing surface waves. The oceans also take in water from the atmosphere as rain, and chemicals such as CO_2, and other trace gases like methane, nitrous oxide, and so on. The surface layers absorb sunlight from the sun, either directly or after scattering by haze or clouds.

In reverse, the oceans react back on the overlying atmosphere, providing sensible heat by conduction, as well as exchanging infrared radiation with the atmosphere. The oceans also evaporate water into the atmosphere, making available latent heat of evaporation, which is often a large term in the total balance of energy transfer between ocean and atmosphere and vice versa, especially in tropical regions.

Thus, these interactions are important for our understanding of how the climate system works, but they are not terribly well understood in quantitative terms. They

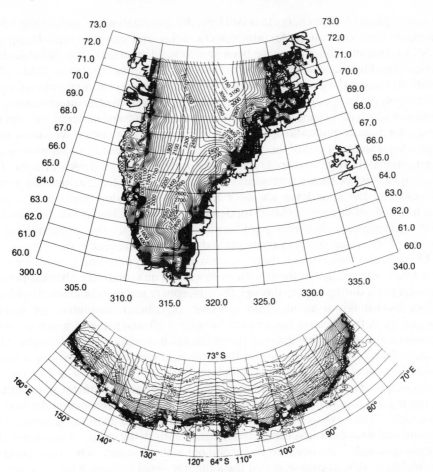

Fig. 5.1 — SEASAT radar altimeter measurements of the topography of Greenland (top) and
Antarctica (bottom).

are processes which occur all over the globe, and which must therefore be understood on a global basis. The study of these ocean—atmosphere interaction processes
is therefore a promising area for the application of satellite remote sensing techniques. A number of different space missions have addressed different aspects of the
problem, but the one mission which most clearly demonstrated the true potential of
space in this area was perhaps SEASAT. This spacecraft carried several radar
systems and demonstrated the power of radar techniques for ocean surface studies
(see Chapter 4). The two examples we will consider here are both taken from
SEASAT results, the first being the measurement of global ocean topography using
the radar altimeter, and the second being the measurement of surface winds from
their effect on the ocean surface through the creation of wind-driven capillary waves,
sensed by the scatterometer on the spacecraft.

As we noted earlier, the altimeter is an instrument which measures the time taken for a short pulse of radar energy to travel from the spacecraft to the surface and back, making various corrections for atmospheric delay and other factors. Using the SEASAT altimeter, scientists have been able to build up maps of the global surface topography of the oceans, which in turn reflect the gravitational potential field and its variations around the globe. Such information is important in accurate modelling of the currents in the upper layer of the oceans, which obviously has direct bearing on the way in which the oceans can transport heat, momentum and materials around the globe. We need to understand such oceanic transport processes if we are to fully comprehend many aspects of the climate system. Plate 15 shows a NASA picture of a measurement of the global topography of the oceans, based on SEASAT data. The topographic surface shown represents the influence of both the gravitational potential field and the underlying sea floor topography, including mid-ocean ridges, trenches and other more isolated topographic features. Clearly, the potential of altimetry measurements is very great for global studies of the oceans even though a number of developments in instruments and techniques are required.

Of particular importance is the separation of the many different terms which have to be taken into account in the equations describing quantitatively what happens to the radar pulse during its path through the atmosphere and in its interaction with the surface (which is not an idealized reflector, of course). Similarly, the surface topography of the ocean is determined by several effects, some of which we have mentioned already: the gravitational potential, and its variation around the globe has an immediate effect, of course, and the motion of the surface waters as currents obviously induces surface gradients in the way described in Chapter 4, as for example can be imagined around eddies, or 'whirlpools'. As we can see from the data in Plate 15, the bottom topography also plays an important role in determining the surface topography. The analysis of the altimeter signal must take all of these factors into account, and since each effect contributes significantly to the nature of the signal, this analysis is complex and needs to be carried out with great care. The corresponding 'up-side' to all this, of course, is that the altimeter signal contains a huge amount of information which is there for our use, as long as we are clever enough to extract it accurately.

The second example of satellite-derived global ocean data comes from the radar scatterometer. The scatterometer, as the name implies, operates by scattering of radar pulses from surface waves with wavelengths close to the actual wavelength of the radiation used. The surface waves of interest in this application are those created directly by the wind as it blows over the surface of the ocean, the so-called capillary waves, with wavelengths typically of a few mm to a few cm. These waves are also known as surface tension waves, since at scales of the order of one cm, the restoring force which is tending to oppose the creation of waves is surface tension. Capillary waves are generated or die away in immediate response to surface winds, as opposed to longer wavelength gravity waves (where the dominant restoring force is gravity) which build up more slowly in response to wind stress (typically over timescales of about one hour). It is an interesting aside to note that the actual mechanisms of transfer of energy from the wind to these latter types of waves are still not firmly established quantitatively, as noted, for example, by Knauss (1978, p. 206). The nature of capillary waves also explains quite simply why a film of oil on the surface of water can cause a 'slick': the surface tension of the oil-covered water is greater than that of clean water, and a slick appears, owing to the fact that capillary waves are less well developed on the polluted water.

The SEASAT scatterometer operated at a frequency of 5.6 GHz, or a wavelength of about 5 cm, that is, close to the wavelength of the wind-driven capillary waves themselves. Thus the signal return strength was dependent upon the magnitude and extent of capillary waves in the radar beam. The radar actually operated at an oblique angle, which meant that the effective wavelength presented by the surface waves to the radar waves was somewhat shortened, or in reverse that the radar wavelength would 'seek out' capillary waves of rather longer wavelength than the radar wavelength itself. However, this is not a first-order effect, and we need only note it in passing. The accuracy of the relationship between the scattering effectiveness of the ocean surface, as measured by the radar scattering coefficient, σ, and the actual wind speed at the surface (as determined by independent measurements at the surface as the spacecraft passed overhead) was found on analysis to be a surprisingly good fit (see Fig. 5.2), despite the fact that so little has been established with any

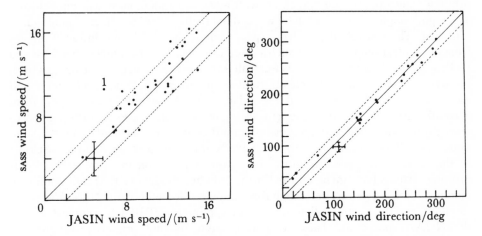

Fig. 5.2 — Scatterometer-derived surface wind speed and direction, compared with surface measurements.

certainty about the physical mechanisms of wave generation and energy transfer processes. On the basis of this demonstration of accuracy of the technique, the global wind fields derived from the SEASAT altimeter results, an example of which is shown in Plate 16, may be taken to be reasonably accurate representations of reality. The ability to accurately measure the global wind stress of the atmosphere on the oceans, which it now appears is within our grasp in future space missions, is obviously of great importance to the monitoring of the climate system.

5.5 SATELLITE OBSERVATIONS OF CLOUDS

The importance of clouds to the global climate was discussed in Chapter 2, where we saw that, despite a great deal of work, some uncertainty exists over the exact role of clouds in the balance of the climate. We found that some work had pointed to a remarkable balance between the net cooling and the net heating effects of clouds, while other work indicated that the overall, global effect was to produce a net cooling, though probably small in magnitude. Considering the vast range in cloud types and properties, and that a typical scene might include a variety of cloud types, at different heights, with differing top temperatures, with a range of emissivities and reflectivities in the visible and the infrared, then it is not surprising, perhaps, that the situation should be uncertain.

Satellite observations are well-matched to the needs of scientists studying clouds and climate. The global nature of the data clearly is needed, since clouds are a global phenomenon, and instruments on satellites can measure quite directly the parameters and variables we need, for example infrared outgoing flux, reflected visible radiation, spectral and integrated albedo. There are difficulties, of course. The accuracy of radiation measurements required is demanding of even the best available sensors. Also, the requirements of global sampling to avoid 'aliasing' of data, and major data gaps are difficult to satisfy without an unrealistic number of separate orbiting vehicles, each instrumented with advanced radiation-measuring sensors; although, sampling schemes can be worked out to minimize the effect of inadequate sampling. The sources of data from space include the visible and infrared images from the geostationary satellites such as Meteosat over Europe, and GOES-East and GOES-West at 75 and 135 degrees west, all three over the equator. The images from these spacecraft provide valuable information on the morphology, development and movement of clouds, especially at low and mid-latitudes where the view aspect from geostationary orbit is favourable, (for example, see Plate 1). Polar orbiting spacecraft also provide data; though these systems do not keep a particular cloud system under surveillance continuously, they provide better global coverage, and being closer to the clouds, are generally more use in direct radiation measurements. Polar orbiting systems have included the Earth Radiation Budget (ERB) radiometer on Nimbus-7, the Advanced Very High Resolution Radiometer on the NOAA weather satellites, and the Earth Radiation Budget Experiment (ERBE) instruments also on the NOAA spacecraft. These various instruments, not primarily designed as imagers, have fairly wide spatial resolution scales, of between 2.5 degrees square to about double this. (Chapter 4 contains further details on some of these instruments and spacecraft systems.)

Plate 17 shows some data for two months in 1979, taken using the NASA Nimbus-7 ERB experiment. This is a record of the monthly averaged three-dimensional distribution of cloud in January and June of that year, with low clouds shown as blue, intermediate clouds as green, and high level clouds as red. This result illustrates the power of satellite data in providing global input data with which to constrain 3-D computer models of the climate. More quantitative measurements, of the net radiative effect of clouds globally, are shown in Fig. 5.3, for the month of April, 1985. The figure shows strong cooling as the lightest shade, moderate cooling as the intermediate shade, and moderate warming as the darkest shade: a moderate global net cooling, which we found in Chapter 2 to represent the current consensus of thinking, is evident from the diagram. As our third example of satellite-derived results Fig. 5.4 represents the zonally averaged (i.e. averaged around a band of latitude) longwave cloud 'forcing', C_{LW}, which we defined in equation (2.47) in

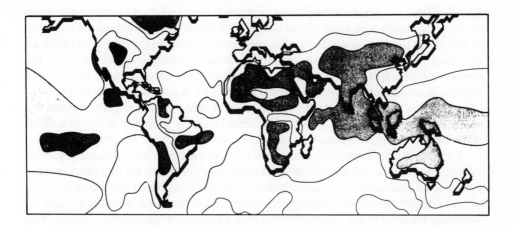

Fig. 5.3 — Effect of clouds on the warming and cooling of the Earth, as recorded by the Earth Radiation Budget Satellite in April 1985: strong cooling, light tint; moderate cooling, medium tint; moderate warming, dark tint.

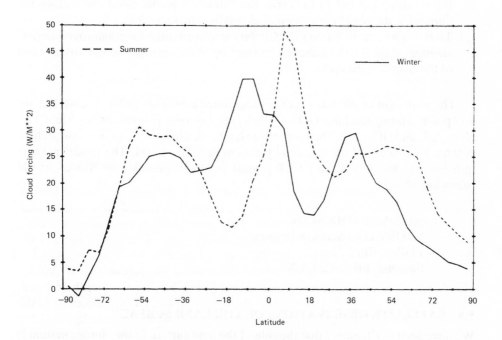

Fig. 5.4 — Longwave cloud forcing (see text for explanation).

Chapter 2. C_{LW} represents the difference in the longwave flux of radiation at the top of the atmosphere with and without clouds. The figure shows the seasonal variation in this cloud forcing parameter, with largest values found in the tropics as might be expected. Minima occur in the generally cloud-free zones associated with the downward branch of the global circulation at latitudes near 20° N and S. It is thought that the maxima in mid-latitudes are associated with storm tracks, and that the low values in polar regions are due to the small difference in temperatures between the surface and the cloud tops, which would mean that the presence or not of a cloud does little to change the radiation properties of that region of the globe.

Clearly, the development of comprehensive and accurate global cloud observing systems from space will be an important component of future climate programmes. A major international programme is already under way, under the auspices of the World Climate Research Programme, sponsored by the International Council of Scientific Unions, ICSU, and the World Meteorological Organization, WMO. This is called the International Satellite Cloud Climatology Project, ISCCP, the aims of which are (see WCP-95, 1985; and Schiffer and Rossow,1985):

(1) To produce a global, reduced-resolution, infrared and visible, calibrated and normalized radiance data set containing basic information on the radiative properties of the atmosphere, from which cloud parameters can be derived.
(2) To stimulate and coordinate basic research on techniques for inferring the physical properties of clouds from the condensed radiance data set and to apply the resulting algorithms to derive and validate a global cloud climatology for improving the parameterization of clouds in climate models.
(3) To promote research using ISCCP data in contributing to an improved understanding of the Earth's radiation budget (top of the atmosphere and surface) and of the hydrological cycle.

The sources of satellite data for this programme are the operational geostationary and polar orbiting satellites GOES, GMS (the Japanese geostationary), Meteosat, NOAA-7, and NOAA-8. The programme is intended to be a 5-year one, and archive centres have been set up in several places around the world. The reader may be interested to know that the ISCCP central archive is provided by NOAA at the following address:

NOAA/NESDIS/NCDC
Satellite Data Services Division
FOB-3, Rm G233
Suitland, MD 20233, USA.

5.6 SATELLITE OBSERVATIONS OF THE LAND SURFACE

We have seen in Chapter 2 that the role of the land surface in the climate system is complex and far from well understood. The gross thermal properties of soil and vegetation mean that land processes are not the most significant ones for the climate in terms of heat storage or movement around the globe; in these respects the atmosphere and the oceans are far more effective. However, the exchange of moisture and chemicals between the land and the air are of first-order importance in climate studies, and indeed, the influence of the land surface on atmospheric energy and momentum fluxes is significant. The remote sensing of the land is not straightforward, however, since the surface is highly variable, ranging from bare sand, through mixed mid-latitude grasslands and woods, to highly vegetated rain forests in the tropics. The variation of emissivity and reflectivity resulting from this diversity is considerable.

Therefore, the thrust of satellite data studies of land surfaces has tended to be less firmly based on theory than for atmospheric or ocean studies, and has tended to be in the direction of empirical investigations of how satellite-derived parameters depend on surface conditions. While one might regret that more emphasis is not placed on fundamental studies of the interactions of land surfaces with electromagnetic radiation, it has to be said that the empirical approach really is the only practical one when faced with the monumental task of making sense of signals received from the land surface.

Earlier, in Chapter 2, we saw that studies of this sort had led to the recognition that in the visible/infrared region, it was shown to be possible to relate the difference in reflected intensity from the surface at two wavelengths on either side of the 'green edge' to the amount of photosynthetically active material present in the surface. In simple terms, this means measuring, in a reasonably quantitative way, the amount of green matter at the surface. We also saw that in the microwave region, the difference between the horizontal and vertical polarizations of the thermally emitted brightness temperature could be related to the amount of growing vegetation at the surface. Equation (2.57) gave the expression for the normalized difference vegetation index, NDVI,

$$\text{NDVI} = \frac{R(\text{NIR}) - R(\text{VIS})}{R(\text{NIR}) + R(\text{VIS})} , \tag{5.1}$$

and the microwave polarization difference temperature, MPDT, may be written as (see equation (2.59)),

$$\text{MPDT} = T(f,\text{hor}) - T(f,\text{vert}) . \tag{5.2}$$

An example of a recent satellite study which has investigated how these two measures of the properties of the surface compare is shown in Plate 18 (a) and (b). This shows both the NDVI (upper frame in each case) and the MPDT (lower frame) for Africa for (a) September 1982 and (b) March 1983, derived from NOAA's AVHRR instrument, and from the NASA SMMR. The authors of the study which used these data (Justice *et al.*, 1989) considered in some detail how these data sets compared with one another. Generally, the two examples follow very similar lines, but there are a number of differences which have been highlighted by the team doing the study. For instance, the areas of Kenya and Tanzania show low values of NDVI, associated with little or no photosynthetic activity, whereas values of MDPT are moderately high. In March, patches of high NDVI activity occur in Sierra Leone, the Ethiopian highlands and the Ivory Coast, which do not show up as isolated regions of activity in the MDPT data.

From their more detailed studies, the authors concluded that the two indices related to different ground phenomona. The NDVI had been shown in previous work to be related to the amount of intercepted (by the beam of the space sensor) photosynthetically active vegetation. However, these data seemed to indicate that this was not so for the MPDT: variations in the MPDT were primarily characterized by the variation in the amount of standing vegetation, with variations being greatest in regions of moderate-to-low standing biomass. This indicated that the signal causing the effect might be the degree of roughness of the vegetation canopy, causing scattering and absorption. These are still very early days in the use of this sort of data for the study of the land surface, but these results indicate that it might be possible to derive quantitatively meaningful results on the properties of a variety of land surface types from remotely sensed data.

An international programme under the auspices of the United Nations Environment Programme, UNEP, has been set up called the International Satellite Land Surface Climatology Project, the aims of which have been described in a recent publication (UNEP: ISLSCP-Report No. 10, Becker *et al.*, 1987). ISLSCP was set up to conduct basic research to improve our understanding of the processes involved in the interactions between the land surface and the atmosphere with the aid of measurements from satellites. It was thought that there would be four components to the programme:

(1) Adaptation of existing soil–vegetation–atmosphere models and their refinement with respect to hydrometeorological and plant–radiation interactions.
(2) Inference of area-averaged physical–biological land surface characteristics from spectral radiances measured by satelliteborne instruments.
(3) Validation and calibration of the methods that are used to extract this information and of the parameterization schemes used in climate models by means of direct measurements at the ground.
(4) Sensitivity studies of interactions between the land surface and climate using climate models in order to narrow down the requirements for type, accuracy and frequency of the satellite observations.

A number of subprojects are included within the ISLSCP programme, for example the First ISLSCP Field Experiment (FIFE), which took place in 1987 in prairie country in Kansas, USA. This project produced a combination of satellite data, airborne radiometer data, heat and moisture flux measurements on site, and biophysical data obtained from on-site physiology experiments. This powerful combination of data allowed a very clear relationship to be derived between the satellite observations (which were from NOAA-9) with the local conditions, and should certainly have improved retrieval algorithms at least for this type of terrain. Another example was the study of a hydrological basin in the south of France in 1986. The chief aim of this campaign was to study the hydrological cycle, and again data were obtained from satellites, aircraft, radars (for precipitation), and surface measurements.

5.7   HIGH RESOLUTION INFRARED SPECTROSCOPY OF THE ATMOSPHERE

For our final example of data obtained from space we will return to the atmosphere (and with the prerogative reserved for the author!) to my own specialism, the spectroscopy of the Earth's atmosphere. The spectroscopy of the atmosphere, especially in the infrared, is important in our study of the climate system for several reasons. First, we have seen how the details of the spectrum, particularly the lines and bands contributed by the so-called 'greenhouse gases', determine the radiative energy balance of the atmosphere, which in turn determines the surface temperature of the atmosphere — perhaps the most direct manifestation of what we sense as our 'climate'. Similarly, these spectral properties also determine the profile of temperature with height in the atmosphere, with all that that means for dynamics and weather. Second, the atmospheric spectrum contains millions of spectral lines which are 'signatures' of the many different chemicals which exist in the atmosphere, some fairly inactive and benign, others either chemically or radiatively very active and important indicators of the state of our climate (e.g. chlorofluorocarbons, carbon dioxide, and ozone). Third, in our exploration of how space techniques can be used to sense surface properties which are important to the climate system, we have found time and again that we must first of all understand the spectral properties of the intervening atmosphere if we are to understand the remotely sensed signals from the surface which are passing through it.

Thus, almost every satellite remote sensing experiment that has been flown in space has either directly or indirectly contributed to our knowledge of the spectral properties of the atmosphere, from the ultraviolet to the microwaves. In many cases, it has been necessary to undertake supporting programmes of study and research to generate the special, detailed spectral information needed by the particular application, by laboratory experiments, by quantum physics calculations or by field experiments.

There have been some space experiments, however, that have been flown specifically to investigate the spectroscopic properties of the atmosphere, and to establish a comprehensive basis of understanding to serve as a foundation for future Earth observation missions. Perhaps the most sophisticated of this type of project, at least for the infrared part of the spectrum, has been the ATMOS experiment. ATMOS was a project funded by NASA, and carried out by C. B. Farmer and his colleagues at the Jet Propulsion Laboratory in California (see Farmer 1987; and Farmer *et al.*, 1987). The JPL team was supported by an international science team which included scientists from Belgium and the UK. The idea behind the project was to fly a very high spectral resolution infrared spectrometer, (known as a Michelson interferometer, after its inventor A. A. Michelson), on the Space Shuttle, and to record the infrared spectrum of the Sun, absorbed by the layers of the atmosphere, as the Shuttle orbited the Earth. The spectral range chosen was from 2 to about 16 microns (5000 to about 600 cm^{-1}), and the maximum spectral resolution that was achievable by the instrument was 0.1 cm^{-1}. Each spectrum, containing 5000–600/0.1

= 44 000 spectral elements, each with about a 10-bit intensity resolution , would be recorded in 2 seconds. That works out to a data rate in excess of 0.5 Mbit/s when associated housekeeping information is included. ATMOS produced data at an alarming rate!

The mode of operation of the experiment was to use the limb sounding technique which was explained in Chapter 4. The Sun was first acquired and then kept locked onto the input aperture of the instrument during sunrise and sunset events by means of a fairly sophisticated Sun tracker under computer control.

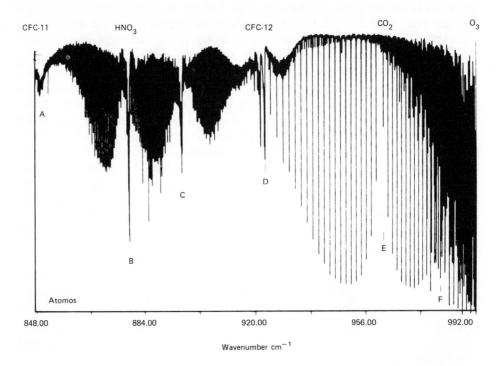

Fig. 5.5 — Results from the ATMOS experiment: the infrared atmospheric 'window' region near 1000 cm^{-1}.

The spectrometer type that was chosen, a Michelson interferometer, was selected because it has a number of advantages over other types of instrument in terms of achievable signal-to-noise ratio, spectral resolution and efficiency in time. When we recall that a spacecraft in polar orbit at Shuttle altitudes is covering the ground at about 7 km/s, efficiency in the use of time is obviously important. The combination of these advantages, coupled with the use of the Sun as a source allowed the designers to achieve the incredible performance figures referred to above. The principal disadvantages of the Michelson interferometer are that it does not produce data directly in terms of a spectrum, but rather the Fourier transform of the spectrum, known as an interferogram: however, with the growth of modern computing power, this has

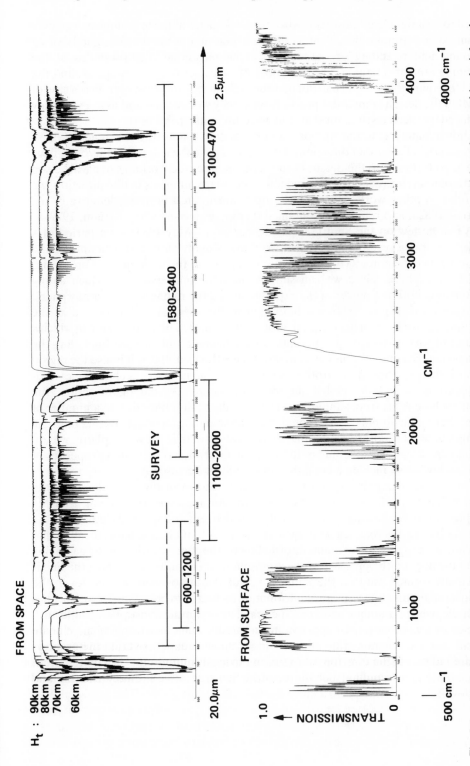

Fig. 5.6 — Synthetic spectra simulating the appearance of the atmosphere from various tangent altitudes from space and from the ground in the infrared wavelength regions covered by the ATMOS instrument. The bandpass regions of the five principal ATMOS filters are shown in the centre of the figure.

ceased to be a significant problem. Also, the advantages of the technique can quickly become signal-to-noise disadvantages if the experiment is not well designed, with a profound understanding of the details of the technique: the subtleties of the technique are considerable, and a number of earlier efforts by groups around the world to exploit the method had not been entirely successful. However, in the case of ATMOS the design team knew precisely what they were doing, and as a result, they produced a superb instrument that fully lived up to its promise.

Unfortunately (for the author — perhaps fortunately for the reader whose interests might lie in other directions!) it would damage the balance of this book if we were to go further into the details of the experiment or the workings of a Michelson interferometer: however, for the reader who does have interests in that direction, we can refer to other works where these topics are treated in more detail (e.g. see Harries, 1982a, and references contained therein, especially Chamberlain, 1979).

ATMOS flew on the Shuttle Challenger in April–May 1985, for a period of a week, and was a resounding success. We have already shown an example of the spectral data in Chapter 2 (Fig 2.8), in our discussion of the basic spectral properties of the atmosphere. Here we can show some additional examples which amply demonstrate both the power of the experiment and the richness of the atmospheric spectrum itself. Fig. 5.5 shows a limb scan in the spectral region known as the atmospheric 'window' in the infrared, around $1000\ cm^{-1}$. This is the region in which a number of surface-sounding experiments have been or will be used (including the Along-Track Scanning Radiometer to fly on the ERS-1 satellite). It is also the region in which the presence of absorption bands of greenhouse gases such as the chloro-fluorocarbons affect the global surface temperature. As the figure shows, the spectrum here is far from simple, and includes absorption lines due to a number of atmospheric constituents. Even though the limb sounding technique exaggerates the strength of weak absorptions, compared to a vertical path in the atmosphere, we can see that any accurate modelling of the spectral absorption or emission properties of this 'window' must include a great deal of detailed spectroscopy.

Our second example of data from ATMOS, shown in Fig. 5.6, illustrates the sheer volume of information contained in the infrared spectrum, being a highly compressed curve showing virtually all of the 40 000 separate spectral elements in the entire 2–16 microns wavelength range covered by ATMOS. When it is recalled that each line in the spectrum contains information on the concentration of the gas giving rise to the line, on the pressure of the atmosphere at the altitude at which the line is principally formed, and on the temperature of the atmosphere at that level, the significance of this huge volume of information can perhaps be grasped. Not only has ATMOS given us a unique source of spectral data about the atmosphere, which will be used in a wide variety of applications in satellite remote sensing of the climate system, but it has demonstrated the enormous care and expertise that are nowadays required to stay at the forefront of exploiting space techniques for climate research. The subject is demanding, but the scientific rewards are extremely exciting and worthwhile.

6

The future — the space station era

6.1 THE FUTURE

In this book we have attempted a synthesis of two aspects of climate research that have to date largely been considered separately. On the one hand, we have tried to give a fairly simple exposition of the scientific basis of our understanding of the climate system, and have spent some time particularily examining some of the issues that most often appear in our newspapers and on radio and television at the present, for example the greenhouse effect and the ozone hole. On the other hand, we have tried to relate these issues to the way in which we can exploit the new technologies and methods of observations of the Earth from space as a tool to improving that understanding, the premise being that the two aspects are inextricably linked, and that in good scientific tradition we cannot treat the science and the technology as quite separate subjects: it is a well demonstrated fact that highest quality science generally results from a very close coupling of theory and of experimental method, the one feeding back on the other, the scientist and engineer having an appreciation of both aspects, even though he or she may not be a specialist in both.

Thus the book has also been presented deliberately to a non-specialist readership, for two reasons. The first is that the rapidly growing concerns with our environment are very much to be encouraged, but they run the risk that our good intentions will quickly outstrip our ability to do something about them, not just in terms of money, but more importantly in terms of the availability of qualified people. We desperately need more young people to enter the field of environmental studies, atmospheric science, oceanography, Earth observation and related areas, equipped with the necessary basic education in physics, mathematics, chemistry, biology, geography and so on that is needed. This book is in part an attempt to interest and enlist those young people. The second reason for the choice of readership 'target' is that for those of us who are already qualified and embarked on our careers, there is a need to recognize the multi-disciplinary nature of climate research, and to cross boundaries in our reading and perhaps in our work: at the introductory level of this book, it may prove a stimulus to such cross-boundary reading, and serve as an introductory or background text for specialists in widely separated disciplines.

Our educational institutions must also rise to the challenge. It is not sufficient to simply adopt the new green shades of politics, nor even to throw money at the problems facing us (although climate research is considerably under-funded still), but flexibility in our educational system is necessary, in order to produce graduates in the required disciplines, and also to run 'conversion courses' so that established scientists can adapt to the new subjects.

As far as the topics for future research are concerned, we might identify several key areas that are going to need much more effort over the coming 5–10 years; going beyond that time is probably a little speculative at this range, considering the major changes of attitudes, of emphasis and of scientific awareness that have taken place in just the last decade. Amongst the more pressing needs are:

- Study of individual processes will need to progress, since it is on the understanding of these processes that our comprehensive models of the climate will build, and which our future observational systems will be designed to investigate and observe. For example, key processes that can even now be listed as requiring urgent attention include: the heterogeneous chemical processes that seem to dominate the photochemistry of the polar night stratosphere; the individual atmosphere–ice–ocean–biosphere feedback processes that we are still only discovering, let alone describing fully; and processes controlling the apparent switching of metastable states in the equatorial Pacific.
- Development of global space systems for remote sensing and monitoring of climate parameters on a continuous and global basis: new instrumental techniques will need to be developed for many problems,(for example, the observation of the troposphere globally from space largely continues to defy us), and new, more powerful methods of assimilating the enormous quantities of data which will be available to us in the future will be required. The coupling of global data sets from space systems to data acquired by more conventional sources, for example by measurements at the surface, will need to be developed further. As a theme to our future work in Earth observations we must keep in mind that even the most sophisticated computer model is inadequate, and contains all manner of limitations, both fundamental and technologically imposed. Therefore, observations are our only contact with reality, and must be made as accurately and comprehensively as we can manage.
- The development of computer models of the climate will also need to continue to advance. At all levels, from simulations of individual processes, through simplified one- and two-dimensional models of the atmosphere, the oceans or the cryosphere, to fully coupled three-dimensional ocean–atmosphere–biosphere models which only the largest and richest institutions worldwide can afford to operate, progress is needed. Also a deeper understanding of the intrinsic randomness of our climate system, and the degree of predictability, is needed to build upon the pioneering work of Lorenz and others, and on the new ideas of chaos theory that have emerged in recent years. Ultimately, the computer models that we as humanity develop incorporate all our knowledge of our planet, and our ability to understand its future. As long as we continually check the results of

model calculations by comparison with real measurements, we must do all we can to improve on those models.

6.2   THE SPACE STATION ERA

What then of the space systems of the future? To date, mankind's study of the climate system from space has been piecemeal, even when considering the major role played by NASA. Missions have been developed largely on a one-off basis, with scientific goals, at least in the past, dictated by the interests of a relatively small band of leading experimental groups around the world. NASA has in the 1980s in particular, however, led the way in developing a more comprehensive, strategic approach to planning its Earth observing programme, and indeed other Agencies around the world, including ESA, ISAS in Japan, and others, have adopted this approach.

In the future, western nations are planning that the space station will be the framework around which space programmes will be planned, and we have seen that within the space station as a whole, it will be the polar platforms (PPF) that will be the most directly applicable to the study of the climate. There are those who argue that the philospophy behind the PPF, and indeed the whole space station compex, that of putting all eggs into one large 'basket', is flawed, and that an approach of flying a larger number of smaller, dedicated free-flyer spacecraft would be more successful. The author must admit to being at one stage amongst this group: however, this is a rather purist approach that to some extent ignores the practicalities of actually getting space missions approved. Space missions are very expensive. They are justified, at least for climate research, as we have seen, by the global nature of the climate problem, and by the huge cost consequences of something 'going wrong' with our climate. Nevertheless, seeking and obtaining funding for a space mission is a risky business, at the mercy of all sorts of pressures and political considerations. The mortality rate in new proposed space missions is therefore very high. Therefore, it may be that the colossal investment in space systems of the future, and in the underpinning research required, may only be realizable by means of developing a space infrastructure programme that can gather the necessary 'momentum' to be realized. This, in some people's view, may be the most powerful justification there is for the space station/PPF programme.

However, we should probably go no further along this track, which is more the province of the political and the social scientist than the physical scientist, and it is the latter discipline to which this book is devoted. Let us turn, therefore, to consider what plans are being made for future Earth observing systems based on the presumption that, for whatever reason, the space station complex, including the PPF, will be a reality.

Both NASA and ESA have been developing their plans for the first few PPFs over recent years, and in 1988 issued a coordinated invitation to the world's scientific community to make proposals for instrumentation and experiments to fly on the first two US platforms, and the first European one. Both agencies have been taking advice from groups of scientists, as well as administrators and politicians, and the proposals received from the scientists were compared by both agencies with guidelines and priorities as laid down by these advisers. The result to date — and this is a 'long-term evolving situation', as an agency official might put it — is that initial payload complements have been chosen by each agency. These payloads are not yet finally determined, but become more so as time goes on, and agency engineers and managers are able to go further with their studies of instrument feasibilities, of payload compatibility and accomodation, of the required ground system, and so on. The payloads are also being determined in what is a reasonably coordinated way: that is to say, the selections are indeed being coordinated, but as yet the agencies are not in total agreement as to what should fly on which platform!

In both the American and the European cases, there has been the definition of a so-called 'core' payload, comprising of instruments which the agency deems are of sufficient general interest from several quarters of the scientific world to merit a central role. Typically, these core instruments are also ones which carry not too great a development risk, i.e. they can be developed without too many nasty surprises appearing to cause delay or cost escalations. The selection of the core payload has also been determined strongly by what data sets need to be measured continuously, building on previous missions, for example global atmospheric temperatures, total ozone amounts, or sea surface temperatures. This obviously raises an interesting question over whether or not the meteorological satellite observations — probably the most universally useful long-term data set in existence — should be carried out in future on the PPFs, or kept as now on separate free-flying satellites. Originally, both the USA and Europe were planning to transfer the meteorological instrument package to the PPFs, but recently (and as some indication of the continuing fluidity in the situation) the USA, in the form of the National Oceanic and Atmospheric Administration, NOAA, has decided to maintain its meteorological observations on free-flyers. Currently, in Europe, ESA and the EUMETSAT organization are considering their position.

In addition, the core payload is intended by both NASA and ESA to be augmented by additional instruments, of a more experimental nature, in many cases carrying greater technical development risk, or representing areas of measurement technique that need considerable development. It is important that agencies maintain a significant proportion of their total payload capability for experimental developments of this type, because (as we have discovered in the preceding pages) we have yet to develop the capability to be able to measure and monitor the full range of parameters that we need, at the levels of accuracy and precision that are required to test our theories of climate properly. In many cases these additional experiments may be more appropriately developed by institutes rather than by agencies and industry, since they may require considerable R&D programmes before they can be realized. In all cases, both core and additional instruments will demand all the effort from all quarters, including industry, institutes, universities, and agencies, if they are to be developed at the level of sophistication, and in the sheer numbers required to fill typical PPF payload capacities of 2000–3000 kg.

It is as yet early in the process of deciding finally on payloads; nevertheless it is perhaps of some value to record in this book what the current plans are. Therefore, in Tables 6.1, 6.2, and 6.3 we note possible instrumental configurations for the American initial missions, though the reader should clearly recognize that many changes are likely before the final complements are settled.

Though it is not perhaps of central interest to us, we should also note that both NASA and ESA have yet to decide on the final design details of the actual spacecraft

Table 6.1 — Possible polar platform PI instrument investigations

ACRIM	**ACTIVE CAVITY RADIOMETER IRRADIANCE MONITOR** **RICHARD C. WILLSON, JPL, Pasadena, California** The objective of ACRIM is to monitor the variability of total solar irradiance with state-of-the-art accuracy and precision. The instrument consists of three total-irradiance detectors. One sensor monitors solar irradiance full time; two sensors calibrate optical degradation of the first sensor.
CERES	**CLOUDS AND THE EARTH'S RADIANT ENERGY SYSTEM** **BRUCE BARKSTROM, LaRC, Hampton, Virginia** The instruments of the CERES investigation will provide Eos with an accurate and consistent data base of radiation. The baseline CERES scanner is two broadband scanning radiometers: one cross-track mode, one rotating plane — similar to ERBE. Each has three channels: total radiance (0.2 to >100 µm), shortwave (0.2 to 3.5 µm), and longwave (6 to 25 µm) as well as a thermistor bolometer detector.
DLS	**DYNAMICS LIMB SOUNDER** **JOHN BARNETT, Oxford University, Oxford, UK** The DLS will measure infrared emission from the stratosphere and mesosphere and obtain high-resolution fields of trace chemicals and temperature. The instrument is a 14-channel infrared limb-scanning radiometer that observes global distribution of upper trophospheric, stratospheric, and mesospheric temperatures and concentrations of O_3, N_2O, CH_4, CFC11, CFC12, and H_2O in the spectral range from 7.04 to 17.06 µm. The resolution is 200 to 400 km east-west and north-south; 3.0 km vertical.
ENAC	**ENERGETIC NEUTRAL ATOM CAMERA** **BARRY H. MAUK, Johns Hopkins, Baltimore, Maryland** ENAC is an imaging particle analyzer assembly that uses Energetic Neutral Atoms (ENAs), as an optical camera uses photons, to remotely image the global structure and dynamics of the Earth's magnetosphere/ionosphere environment. ENAs and charged ions are separetely analyzed as to their energies (20 keV to several MeV), mass composition (e.g. H, He, CNO, etc.), and arrival direction (3°×3° pixels).
EOSP	**EARTH OBSERVING SCANNING POLARIMETER** **LARRY D. TRAVIS, GSFC/GISS, New York, New York** EOSP is a photopolarimeter providing radiance and linear polarization degree in 12 spectral bands from 410 to 2,250 nm. A cross-track scanning mirror sweeps the 10 km field of view (at nadir) from limb to limb, generating global maps. EOSP will determine cloud properties, aerosol distribution, atmospheric corrections, and land and vegetation characteristics.
GGI	**GPS GEOSCIENCE INSTRUMENT** **WILLIAM G. MELBOURNE, JPL Pasadena, California** GGI is a high-performance Global Positioning System (GPS) receiver-processor that includes 18 dual-frequency satellite channels and three distributed GPS antennas. GGI will serve four principal science objectives: centimeter-level global geodesy; high-precision atmospheric-temperature profiling; ionospheric gravity-wave detection and tomographic mapping; and precise position and attitude determination in support of other science instruments.
GOS	**GEOMAGNETIC OBSERVING SYSTEM** **ROBERT LANGEL, III, GSFC, Greenbelt, Maryland** GOS will measure the magnetic field using a three-axis fluxgate and a scalar helium magnetometer. The data will be used to study the Earth's interior and electrodynamic ionosphere coupling. Models of the Earth's field will be generated and used to study such things as the underlying core fluid dynamics and correlations with length-of-day changes and mantle conductivity. A boom-mounted vector fluxgate, scalar helium magnetometer, and three star trackers are employed to obtain absolute scalar (±1 nT) and high accuracy vector (±3 nT, per axis, rss) magnetic-field measurements referenced to a geocentric inertial coordinate system.
HIMSS	**HIGH-RESOLUTION MICROWAVE SPECTROMETER SOUNDER** **ROY W. SPENCER, MSFC, Huntsville, Alabama** HIMSS is a high-efficiency passive microwave radiometer for measurement of numerous atmospheric and oceanic parameters (see instrument measurements matrix). Multifrequency (6.6 to 90 GHz) dual polarization data are acquired with a mechanically scanned 2-meter antenna providing a 1500 km swath width and a constant earth incidence angle of 53°. Radiometric sensitivity varies from 0.25K at 6.6GHz to 0.7K at 90 GHz.

Table 6.1 — Continued

HIRRLS	**HIGH-RESOLUTION RESEARCH LIMB SOUNDER** **JOHN GILLE, NCAR, Boulder, Colorado** HIRRLS is a multi-channel limb scanning infrared radiometer that determines temperature, O_3, H_2O, CH_4, N_2O, NO_2, HNO_3, and other species with a vertical resolution of 2 km from cloud tops to the mesopause. The instrument scan pattern is flexible and controllable from the ground. In the global mode, horizontal resolution is 4° latitude and longitude, allowing small scale dynamical and transport phenomena to be observed.
IPEI	**IONOSPHERIC PLASMA AND ELECTRODYNAMICS INSTRUMENT** **RODERICK HEELIS, UT Dallas Center for Space Sciences, Richardson, Texas** THE IPEI instrument will be used to determine magnetospheric input to the atmospheric and electric fields resulting from neutral wind motion. Information derived from IPEI will be used in conjunction with energetic particles and magnetometer data to determine energy input to the lower atmosphere from high latitudes expressed in terms of the poynting flux. The instrument consists of a retarding mass analyzer and ion drift meter.
LIS	**LIGHTNING IMAGING SENSOR** **HUGH CHRISTIAN, MSFC, Huntsville, Alabama** The calibrated optical LIS will acquire and investigate the distribution and variability of lightning over the Earth. LIS consists of a staring telescope/filter imaging system, which provides 90% detection efficiency under both day and night conditions using a background remover and event processor. It will provide storm-scale (10 km) spatial resolution with 1 ms temporal resolution.
MLS	**MICROWAVE LIMB SOUNDER** **JOE W. WATERS, JPL, Pasadena, California** The MLS investigation will study and monitor processes that govern stratospheric and mesospheric ozone, with emphasis on potential ozone depletion by mankind's increasing industrial activities. The instrument consists of a passive microwave limb-sounding radiometer that obtains vertical profiles of all molecules and radicals believed to be involved in the ozone destruction cycle. Measurements are in the three spectral bands, 637, 560, and 205 GHz, with spatial resolution of 100 km×3 km×6 km.
MISR	**MULTI-ANGLE IMAGING SPECTRORADIOMETER** **DAVID J. DINER, JPL, Padadena, California** The MISR experiment addresses the study of the climatic and environmental effects of atmospheric aerosols, climatic effects of cloud-field heterogeneities, vegetation-atmosphere interactions as they relate to radiation budget, ecosystem changes resulting from natural and anthropogenic activities, and phytoplankton pigment concentration in the tropical oceans. The instrument consists of eight identical CCD-based cameras at four viewing singles 28.5°, 45.6°, 60°, and 72.5°, fore and aft, providing continuous simultaneous imaging in four narrow spectral bands from 440 nm to 860 nm.
MOPITT	**MEASUREMENTS OF POLLUTION IN THE TROPOSHERE** **JAMES DRUMMOND, University of Toronto, Toronto, Canada** MOPITT is a correlation radiometer using pressure modulation and length modulation techniques to measure carbon monoxide (CO) concentrations in the troposphere with the primary objective of enhancing knowledge of the lower atmosphere system and particularly how it interacts with the surface, ocean, and biomass systems. MOPITT measures upwelling infrared radiation in the CO bands at 4.7 μm and 2.4 μm. Using knowledge of the atmospheric temperature profile, global distributions of CO can be determined with limited height profiles. Profiles with a resolution of 25 km horizontally and 3 km vertically, with an accuracy of 10%, will be obtained throughout the troposphere.
POEMS	**POSITRON ELECTRON MAGNET SPECTROMETER** **PAUL A. EVENSON, Bartol Research Institute, University of Delaware, Newark, Delaware** POEMS uses a solid state hodoscope, permanent magnet, and other detectors to measure time dependent energy spectra of these particles at energies greater than 5 MeV. Solar flare neutrons and gamma rays will also be measured. POEMS will measure the critical positron (e^+) and electron (e^-) components of cosmic radiation and utilize this information to trace processes occurring in solar flares in the heliosphere, within our geospace environment, and elsewhere in the galaxy.
SAFIRE	**SPECTROSCOPY OF THE ATMOSPHERE USING FAR INFRARED EMISSION** **JAMES M. RUSSELL, III, LaRC, Hampton, Virginia** SAFIRE is a multichannel far-infrared Fourier transform spectrometer (0.004 wave number spectral resolution) and LIMS-type filter radiometer for measuring trace species in the upper atmosphere. The instrument views the Earth's limb and detects the thermal emission of rotational and vibrational-rotational features using a 48-element far-infrared array in the 80 to 400 wave-number range and a 105-element array in the 630 to 1560 wave-number range.

Table 6.1 — Continued

SAGE III	**STRATOSPHERIC AEROSOL AND GAS EXPERIMENT III** **M. PATRICK McCORMICK, LaRC, Hampton, Virginia** SAGE III is an Earth-limb scanning grating spectrometer for the measurement of O_3, aerosols, H_2O, NO_2, clouds, and air density profiles with 1 km vertical resolution from the mid-troposphere through the lower mesosphere. Spectral channels cover from 300 nm to 1,500 nm. The instrument is self-calibrating, providing the capability of measuring long-term trends in the above species. It is also important for removing cloud and aerosol effects from the data of other remote sensors.
SCANSCAT	**ADVANCED SCATTEROMETER FOR STUDIES IN METEOROLOGY AND OCEANOGRAPHY** **MICHAEL H. FREILICH, JPL, Pasadena, California** SCANSCAT is an advanced, scanning pencil-beam scatterometer, operating at 13.995 GHz and capable of acquiring accurate, high-resolution (25 km), all-weather measurements of surface wind speed and direction over the tropical oceans and elsewhere. SCANSCAT provides wind speed accuracies of 20% at less than 3 m/s and 10% for 3 to 30 m/s, directional accuracy of 20% with 25-km spatial resolution, and coverage over a 1,100 km swath.
SOLSTICE	**SOLAR STELLAR IRRADIANCE COMPARISON EXPERIMENT** **GARY ROTTMAN, University of Colorado, Boulder, Colorado** SOLSTICE provides precise daily measurements of the full disk solar ultraviolet irradiance between 4 to 440 nm. The SOLSTICE instrument consists of a four-channel spectrometer together with the required gimbal drive to point SOLSTICE at the sun and stellar targets.
SWIRLS	**STRATOSPHERIC WIND INFRARED LIMB SOUNDER** **DANIEL J. McCLEESE, JPL, Pasadena, California** The SWIRLS investigation will describe stratospheric structure, dynamics, and image measurements of global irradiances in the 3 to 5 μm and 8 to 13 μm atmospheric windows. The instrument acquires continuous vertical profiles of horizontal wind vectors, temperatures, and pressures and mixing ratios of ozone and nitrous oxide. Wind velocities are determined from wind-induced Doppler shifts in the N_2O emission spectrum. The instrument consists of a gas-correlation and filter-infrared radiometer, which observes atmospheric infrared emissions in the 7.6 μm to 17.2 μm range. It provides 3 km vertical resolution in a 20 km altitude range on both the day and night sides of Earth.
TES	**TROPOSPHERIC EMISSION SPECTROMETER** **REINHARD BEER, JPL, Pasadena, California** TES is a passive, high-spectral-resolution, cryogenic, infrared, imaging Fourier transform spectrometer with sufficient spectral coverage to permit the near-simultaneous measurement of most infrared-active minor constituents of the troposphere. The instrument has two operating modes: nadir mode, offering good horizontal spatial resolution, and limb-viewing mode offering enhanced sensitivity for trace constituents with good vertical resolution, but limited horizontal resolution.
TIGER	**THERMAL INFRARED GROUND EMISSION RADIOMETER** **ANNE B. KAHLE, JPL, Pasadena, California** TIGER will provide quantitative spectral measurements of the emitted radiation from the Earth's surface at spatial and spectral resolutions appropriate for geological, climatic, hydrological, and agricultural studies. The instrument utilizes a Thermal Infrared Mapping Spectrometer (TIMS), which will have a total of 14 channels. It provides spectral image measurements of global irradiances in the 3 to 5 μm and 8 to 13 μm atmospheric windows. The instrument has an IFOV of 90m and a swath of 30.2 km.
TRACER	**TROPOSPHERIC RADIOMETER FOR ATMOSPHERIC CHEMISTRY AND ENVIRONMENTAL RESEARCH** **HENRY G. REICHLE, JR., LaRC, Hampton, Virginia** TRACER is designed to measure the global distribution of CO at multiple levels in the troposphere and, thus, provide a global data base for modeling studies that will increase understanding of global tropospheric chemistry and transport process, as well as CO mixing ratios and geographical distribution within layers versus time. The instrument is a nadir-viewing gas-filter radiometer operating in the 2.3 μm and 4.6 μm spectral regions.
XIE	**X-RAY IMAGING EXPERIMENT** **GEORGE K. PARKS, University of Washington, Seattle, Washington** XIE is a multiple instrument assembly for local and global monitoring of energy input to the Earth's atmosphere by charged particles. The assembly includes particle energy spectrometers and an X-ray imaging spectrometer. The X-ray instrument system consists of a proportional gas-filled counter for detecting 3 to 20 keV X-rays and an Anger Camera for detecting and imaging 20 to 200 keV X-rays.

These instruments represent the current selection

Table 6.2 — The planned international polar platform programme

NASA POLAR PLATFORM-1 NPOP-1	NASA POLAR PLATFORM-2 NPOP-2	ESA POLAR PLATFORM-1 EPOP-1/A1	ESA POLAR PLATFORM-2 EPOP-2/B1	JAPANESE POLAR PLATFORM JPOP	NOAA FREE-FLYER	ATTACHED PAYLOADS
Orbit: 705 km	Orbit: 705 km	Orbit: 824 km	Orbit: 705 km	Orbit: 800 km	Orbit: 824 km	Orbit: 400 km
Crossing: 1:30 pm	Crossing: 1:30 pm	Crossing: 10:00–10:30 am	Crossing: 10:00–10:30 am		Crossing: 1:30 pm	Crossing: 28.5°
Launch: 4th Qtr. 1996	Launch: 4th Qtr. 1998	Launch: 1997	Launch: 2000	Launch: 4th Qtr. 1998	Launch: 1st Qtr. 1998	Launch: 1st Qtr. 1998
AIRS	SAR	ATLID	ATLID	LAWS	ARGOS	MODIS-N
ALT	SEM	MERIS	HRIS	AMSR	AMRIR	MODIS-T
GLRS	GGI	MIMR	MIMR	AVNIR	AMSU	MIMR
HIRIS	GOS	AMIR	SAR-C	OCTS	GOMR	AMSR-2
MODIS-N	IPEI	CHEMISTRY	AMIR	SAR-L	SEARCH & RESCUE	ERBI
MODIS-T	LIS	RADIOMETER	CHEMISTRY	SAR-X	SEM	OZONE SENSOR
SEM	MLS	ALTIMETER (GPS)	STEREO IMAGER			PPS-PODS
MIMR	SAFIRE	AMI-2				RAIN RADAR
AMSR	SWIRLS	AMRIR				SCATTEROMETER
ITIR (TIGER)	TES	AMSU				SPECIAL IMAGER
CERES	XIE	SEM				SOLAR FLIGHT
DLS		ARGOS				OR OPPORTUNITY
ENAC		SEARCH &				1st QUARTER 1995
EOSP		RESCUE				ACRIM
GGI						SOLSTICE
HIMSS						
HIRRLS						
IPEI						
MISR						
MOPITT						
POEMS						
SAGE III						
SCANSCAT						
TRACER						
AMSU						

Fig. 5.6 — Synthetic spectra simulating the appearance of the atmosphere from various tangent altitudes from space and from the ground in the infrared wavelength regions covered by the ATMOS instrument. The bandpass regions of the five principal ATMOS filters are shown in the centre of the figure.

Table 6.3 — Polar platform sensors and measurement capabilities

RESEARCH FACILITY INSTRUMENTS

Instrument	1	2	3	4	5	6	7	8	9	10	11	12	13	14	15	16	17	18	19	20	21	22	23	24	25	26	27	28
AIRS								•	•							•	•	•	•	•							•	
ALT									•	•	•																	
AMSR								•	•	•	•	•				•												
ATLID							•								•													
GLRS	•						•		•																			
HIRIS					•	•	•		•							•								•				
HRIS					•	•										•												
ITIR					•	•		•																				
LAWS						•		•																				
MIMR					•				•		•							•										
MODIS-N						•		•	•	•				•	•	•		•									•	
MODIS-T					•	•		•	•					•	•	•		•									•	
SAR					•	•	•	•	•	•																		
SEM																											•	

OPERATIONAL FACILITY INSTRUMENTS

Instrument	1	2	3	4	5	6	7	8	9	10	11	12	13	14	15	16	17	18	19	20	21	22	23	24	25	26	27	28
AMRIR								•	•	•						•	•	•										
AMSU A/B								•	•		•								•	•	•							
GOMR																							•	•			•	

PI INSTRUMENT INVESTIGATIONS

Instrument	1	2	3	4	5	6	7	8	9	10	11	12	13	14	15	16	17	18	19	20	21	22	23	24	25	26	27	28
ACRIM																												•
CERES																		•									•	
DLS															•		•		•				•	•				
ENAC																									•			
EOSP															•			•									•	
GGI	•		•																									
GOS																										•	•	
HIMSS								•	•	•	•					•	•	•			•							
HIRRLS															•				•	•				•				
IPEI																									•	•		
LIS																						•						
MLS																				•				•				
MISR						•									•			•									•	
MOPITT																	•											
POEMS																									•			
SAFIRE																			•					•				
SAGE III															•	•		•	•	•				•				
SCANSCAT							•		•		•																	
SOLSTICE																												•
SWIRLS																	•			•			•					
TES																•	•		•	•				•				
TIGER					•	•	•																					
TRACER																	•											
XIE																									•	•		

Key:
1: Orbit determination; 2: Interior earth structure; 3: Plate motion and crustal deformation; 4: Surface soil moisture and wetlands extent; 5: Land surface composition; 6: Land surface biological activity, phenology, physiology; 7: Surface topography; 8: Surface temperature; 9: Snow and ice extent and character; 10: Sea ice extent, character, and motion; 11: Sea surface winds; 12: Ocean waves; 13: Ocean circulation; 14: Oceans and lakes biological activity; 15: Aerosols; 16: Tropospheric composition; 17: Tropospheric winds; 18: Cloud properties; 19: Atmospheric temperature; 20: Atmospheric water content; 21: Precipitation rate; 22: Lightning; 23: Upper atmospheric winds; 24: Upper atmospheric composition; 25: Particles and fields environment; 26: Ionosphere; 27: Earth radiative balance; 28: Solar output.

Notice that some instruments make multiple measurements (indicated by rows of dots). In addition, several different instruments are needed, in most cases, to satisfy requirements within one measurement category (indicated by columns of dots).

to be used as the 'buses' for the PPFs. Indeed, within ESA there has been considerable controversy over the choice of bus, with designs being put forward by both France and the UK, and only after many months of negotiation a compromise solution emerging. Bearing in mind that these words are being written early in 1990, this calls into question whether we are actually going to see a PPF in orbit by 1997 or so, as is being forecast currently, or whether more realistically we should look forward to the event as a New Year present at the beginning of the third millennium. For the doubters, Plate 19 shows an artist's impression of a NASA design for the PPF.

Planning for the future Earth observing systems is not restricted to the USA or Europe, of course. Most notably, Japan is planning a major contribution to the manned space station, and has strong interests in polar orbiting missions of its own, including the contribution of instruments to the US PPFs. Canada is planning its own synthetic aperture radar satellite project, Radarsat, with an eye on its unique problems of exploiting the cloudy ice-bound polar regions of northern Canada and the north-western Atlantic. The USSR, of course, has its own, well-established space station programme, though this seems more concerned with the problems of manned spaceflight and the exploration and exploitation of space itself than the study of the Earth's climate from space. Certainly, the Soviet Union does have polar orbiting Earth observation projects of its own, but from what we in the west can gather, the instruments flown are either not particularly advanced, or are too advanced to be revealed to the non-military community. Notwithstanding this, it is very much to be hoped that the USSR, with its considerable resources and intellectual capabilities, will participate fully in future climate research programmes using space.

6.3   FINAL REMARKS

Let us conclude this chapter with some remarks on how these future plans and programmes that we have outlined should develop, with all due apologies from the author for the occasionally unjustified and subjective view. The climate problem is probably the biggest single problem to face mankind today, aside from the hopefully diminishing possibility of global thermonuclear war. As such, nations have to cooperate in the understanding of how the climate system works, and what might become of it, through natural and man-made causes in the future. Nations must not stint in providing the resources necessary for this task, nor in educating the young people required to carry it through. After all, an investment equal to a small fraction of the essentially wasted expenditure on the arms race, and on all other aspects of war, would be more than enough.

The space and environmental agencies are facing up to their responsibilities well, and programmes such as the space station, as we have seen, are actively being developed. However, there are dangers, difficulties and things to be aware of. Thus, the design of the space systems that are to be developed should be inextricably linked to the requirements of the science: the selections of orbits, of instruments and so on should be driven directly by the problems we are trying to solve. While this book has paid due tribute to the efforts of the space agencies, and particularly NASA, we must sound a note of caution here. For it is still true that the selection of experiments or instruments, while increasingly guided by the scientific requirements, is unduly influenced, at least in this author's view, by the politics and the power games that are played out by the major institutions, for whom the need to achieve selection in 'the next' instrument opportunity is intense, and understandable.

The programmes are also in danger, some people would hold, of becoming too massive, and too unwieldy, to really respond to the needs of mankind in facing up to the climate issue. While we argued above that the necessary investment in future space programmes may only be realizable by going the route of large projects like space station, programme managers must ensure that things do not get out of hand. A number of basic goals need to be kept in mind: space needs to be accessible, meaning without too great a cost, and without great gaps between flight opportunities; the Earth observing system needs to provide highly accurate long-term databases on the components of the climate system; and the Earth observing system also needs to be responsive and flexible, so that new phenomena and climatic problems can be quickly assessed and monitored; and finally, the investment in the space segment of new programmes needs to be matched by a sufficient investment in the data systems, the information networks, and of course the people to make sense of it all. While there are obvious risks that programmes will become unwieldy, too great for the human scale of the endeavour to be retained, there is much evidence from the excellent projects and plans of NASA, ESA and the other agencies involved that these dangers are fully recognized, and that programmes will be wisely managed.

So we come to the end of this particular story. The challenge of the climate problem is a major one, but one we are capable of meeting if we organize ourselves well. The scientific issues are fascinating, and provide an enormously difficult test of our scientists; the space systems which we need to help in our understanding of those issues will require no less ingenuity of our engineers, scientists and managers. Ultimately the object of our attention, the Earth on which we live, is not only an essential system for our survival, it is also a very beautiful subject which any scientist would be grateful to have as a focus of his work.

References

Allan, T. D. (1983), *Satellite Microwave Remote Sensing*, Ellis Horwood Ltd, Chichester.

Ardanuy, P. E., Stowe, L. L., Gruber, A., Weiss, M. and Long C. S. (1989), *J. Climate*, **2** 766–799.

Arking, A. (1989), *Status of knowledge of the radiative effects of clouds and their impact on climate*, Report to the International Radiation Commission, July 1989.

Atkinson, R. J., Matthews, W. A., Newman, P. A. and Plumb, R. A. (1989), *Nature* **340** 290–294.

Arrhenius, S. (1896), *Phil. Mag.* **41** 237–276.

Barton, I. J. and Cechet, R. P. (1989), *J. Atm. Ocean. Tech.* **6** 1083–1089.

Becker, F., Bolle, H. J. and Rowntree, P. (1987). 'The International Satellite Land Surface Climatology Project', ISLSCP Report No. 10.

Bigg, G. R. (1990), *Weather* **45** 2–8.

Bjerknes, J. (1969) *Mon. Weather Rev.* **97** 163–172.

Bolin, B. (1984), 'Biogeochemical processes and climate modelling', in Houghton (Ed.) 1984.

Born, M. and Wolf, E. (1975) *Principles of Optics*, Pergamon Press.

Brasseur, G. and Solomon, S. (1984), *Aeronomy of the Middle Atmosphere*, D. Reidel and Co.

Budyko, M. I. (1969), *Tellus,* **21** 611–619.

Budyko, M. I. (1974) *Climate and Life,* Academic Press.

Callendar, G. S. (1938), *Q. J. Roy. Met. Soc.* **64** 223–240.

Cess, R. D. (1974), *J. Quant. Spec. Rad. Trans.* **14** 861–872.

Cess, R. D. (1976), *J. Atmos. Sci.* **33** 1831–1843.

Cess, R. D., Briegleb, B. P. and Lian, M. S. (1982), *J. Atmos. Sci.* **39** 53–59.

Chahine, M. T., Haskins, R., Susskind, J. and Reuter, D. (1986), *Proc. ISLSCP Conference,* Rome, Italy, Dec. 1985, ESA SP-248.

Chamberlain, J. (1979), *The Principles of Interferometric Spectroscopy,* John Wiley.

Chapman, S. (1930), *Mem. R. Met. Soc.* **3** 103.

Choudhury, B. J. (1989), *Int. J. Rem. Sensing* **10** 1579–1605.

Coakley, J. A. and Wielicki, B. A. (1979), *J. Atmos. Sci.* **36** 2031–2039.

Crutzen, P. J. (1970), *Q. J. R. Met. Soc.* **96** 320–325.

Douglass, A. R. and Stanford, J. L. (1982), *J. Geophys. Res.* **87(C7)** 5001–5008.

Edwards, T., Browning, R., Delderfield, J., Lee, D. J., Lidiard, K., Milborrow, R. S., McPherson, P. H., Peskett, S. C., Toplis, G. M., Taylor, H. S., Mason, I., Mason, G., Smith, A. and Stringer, S. (1990), *J. Brit. Int. Soc.* **43** 160–180.

Farman, J. C., Gardiner, B. G. and Shanklin, J. D. (1985), *Nature* **315** 207–209.

Farmer, C. B. (1987), *Mikrochim. Acta* **3** 189–214.

Farmer, C. B ., Raper, O. D. and O'Callaghan, F. G. (1987), 'Final report on the first flight of the ATMOS instrument during the Spacelab 3 mission, April 29 through May 6 1985', JPL Publication 87–32.

Gates, W. L., Cook, K. H. and Schlesinger, M. (1981), *J. Geophys. Res.* **86(C7)** 6385–6393.

Gates, W. L., Han, Y. J. and Schlesinger, M. E. (1985), in *Coupled Ocean–Atmosphere Models,* (Ed.) Nihoul J. C. J., Elsevier Oceanography Series, Vol. 40, 131–151.

Gill, A. E. (1980), *Q. J. R. Met. Soc.,* **106** 447–468.

Gill, A. E. (1982), *Atmosphere–Ocean Dynamics,* Academic Press.

Gill, A. E. and Rasmussen, E. M. (1983), *Nature* **306** 229–234.

Gille, J. C. and House, F. B. (1971), *J. Atmos. Sci.* **28** 1427–1442.

Gille, J. C. and Russell, J. M. (1984), *J. Geophys. Res.* **89(D4)** 5125–5140.

Gille, J. C., Russell, J. M., Bailey, P. L., Remsberg, E. E., Gordley, L. L., Evans, W. F. J., Fischer, H., Gandrud, B. W., Girard, A., Harries, J. E. and Beck, S. A. (1984) *J. Geophys. Res.* **84(D4)** 5179–5190.

Gleick, J. (1987), *Chaos,* Sphere Books.

Goody, R. M. (1964), *Atmospheric Radiation,* Oxford University Press.

Gornitz, V., Lebedeff, S. and Hansen, J. (1982), *Science* **215** 1611–1614.

Guymer, T. H. (1983), *Phil. Trans. R. Soc. Lond.* **A309** 399–414.

Hansen, J. E., Johnson, D., Lacis, A., Lebedeff, S., Lee, P., Rind, D. and Russel, G. (1981), *Science* **213** 957–966.

Hansen, J., Fung, I., Lacis, A., Rind, D., Lebedeff, S., Ruedy, R., Russell, G. and Stone, P. (1988), *J. Geophys. Res.* **93** 9341–9364.

Hanson, K., Brier, G. W. and Maul, G. A. (1989), *Geophys. Res. Letts.* **16** 1181–1184.

Hardisky, M. M., Gross, M. E. and Klemas, V. (1986), *Bioscience* **36** 453–460.

Harries, J. E. (1982a), 'Infrared and submillimetre spectroscopy of the atmosphere', in *Infrared and Millimeter Waves,* Vol. 6, Academic Press.

Harries, J. E. (1982b), *J. Atm. Terr. Phys.* **44** 591–597.

Harries, J. E., Llewellyn-Jones, D. T., Minnett, P. J., Saunders, R. W. and Zavody, A. M. (1983), *Phil. Trans. R. Soc. Lond.* **A309** 381–395.

Hartman, D. L. and Short, D. A. (1980), *J. Atmos. Sci.* **38** 1233–1250.

Hastenrath, S. (1988), *Climate and Circulation of the Tropics,* D. Reidel.

Henderson-Sellers, A., Wilson, M. F., Thomas, G. and Dickinson, R. E. (1986), 'Current global land surface data sets for use in climate related studies', National Center for Atmospheric Research, Boulder, Report NCAR/TN-272 STR.

Hoskins, B. J. and Simmons, A. J. (1975), *Q. J. R. Met. Soc.,* **101** 637–655.

Hoskins, B. J., James, I. and White, G. H. (1983), *J. Atmos. Sci.*, **40** 1595–1612.

Houghton, J. T. (1990), (Chairman), 'Scientific Assessment of Climate Change', Report of Working Group 1 of the Intergovernmental Panel on Climate Change (IPCC), World Meteorological Organization and United Nations Environmental Programme.

Houghton, J. T. (1977) *The Physics of Atmospheres,* Cambridge University Press.

Houghton, J. T. (Ed.) (1984), *The Global Climate*, Cambridge University Press.

Houghton, J. T. and Morel, P. (1984), 'The World Climate Research Programme', in Houghton (Ed.) (1984).

Houghton, J. T. and Smith, S. D. (1966), *Infrared Physics*, Oxford University Press.

Houghton, J. T., Taylor, F. W. and Rodgers, C. D. (1984), *Remote Sounding of Atmospheres*, Cambridge University Press.

Hoyt, D. V. (1976), NOAA Technical Report, ERL 362-ARL4, 1–124.

Ingram, W. J., Wilson, C. A. and Mitchell, J. F. B., (1989), *J. Geophys. Res.*, **94**(D6) 8609–8622.

Jones, P. D., Wigley, T. M. L. and Wright, P. B. (1986) *Nature* **322** 430–434.

Jones, R. L., Pyle, J. A., Harries, J. E., Zavody, A. M., Russell, J. M. and Gille, J. C. (1986), *Quart. J. R. Met. Soc.* **112** 1127–1143.

Justice, C. O., Townshend, J. R. G. and Choudhury, B. J. (1989), *Int. J. Rem. Sensing* **10** 1607–1632.

Knauss, J. A. (1978), *Introduction to Physical Oceanography,* Prentice-Hall, USA.

Kondratyev, K. Ya., (1969), *Radiation in the Atmosphere,* Academic Press.

Kuo, C., Linberg, C. and Thomson, D. J. (1990), *Nature* **343** 709–714.

Leith, C. E., (1983) 'Global Climate Research' in *The Global Climate,* Ed. J. T. Houghton (1984).

Lorenz, E. N. (1963a), *J. Atmos. Sci.* **20** 130–141.

Lorenz, E. N. (1963b), *J. Atmos. Sci.* **20** 448–464.

Lorenz, E. N. (1968), *Meteorol. Monographs* **8** 1–3.

Manabe, S. (1983), 'Carbon Dioxide and Climate Change', in Saltzman (Ed.) (1983).

Manabe, S. and Stouffer, R. J. (1979), *Nature* **282** 491–493.

Manabe, S. and Stouffer, R. J. (1980), *J. Geophys. Res.* **85**(C10) 5529–5554.

Manabe, S. and Stouffer, R. J. (1988), *J. Climate,* **1** 841–866.

Manabe, S. and Wetherald, R. T. (1967), *J. Atmos. Sci.,* **24** 241–259.

Manabe, S. and Bryan, K. (1969), Monthly Weather Review, **97** 739–827 (three consecutive papers).

Manabe, S. and Wetherald, R. T. (1980), *J. Atmos. Sci.* **37** 99–118.

Mason, B. J. (1974), *The Physics of Clouds,* Oxford University Press.

McCormick, M. P.,, Zawodny, J. M., Veiga, R. E., Larsen, J. C. and Wang, P. H. (1989), *Plan. Space Sci.* **37** 1567–1586.

Meeks, M. L. (Ed.) (1976), *Methods of Experimental Physics,* Vol. 12, Academic Press.

Meier, M. F. (1983), *J. Hydrolog. Sci.* **28** 3–22.

Mintz, Y. (1984), 'The sensitivity of numerically simulated climate to land surface boundary conditions', in Houghton (Ed.) (1984) Cambridge University Press.

Mitchell, J. F. B. (1979), Met. Office Techn. Note No. II/137, Meteorological Office, Bracknell, UK.

NASA (1985), *The Stratosphere 1985,* NASA, Washington DC.

NASA (1988a), 'Earth system science: a closer view', Report of the Earth System Sciences Committee, NASA Advisory Council, Washington DC.

NASA (1988b), 'From pattern to process: the strategy of the Earth Observing System', EOS Science Steering Committee Report, volume II, NASA, Washington DC.

NASA (1988c), Reference Publication 1208, 'Present state of knowledge of the upper atmosphere: an assessment report', NASA, Washington DC.

National Academy of the US (1986), *Global Change in the Geosphere–Biosphere,* National Academy Press.

Ohring, G. and Adler, S. (1978), *J. Atmos. Sci.* **35** 186–205.

Ohring, G. and Gruber, A. (1983), 'Radiation Observations and Climate Theory', in Saltzman (Ed.) (1983).

Ohring, G., Clapp, P. F., Heddinghaus, T. R. and Kreuger, A. F. (1981), *J. Atmos. Sci.* **38** 2539–2541.

Peckham, G. E., Harwood, R. S., Kerridge, B. J., Matheson, D. N. and Farman, J. C. (1988), 'EOS Microwave Limb Sounder: UK proposal to the British National Space Centre' (available from Rutherford Appleton Laboratory).

Pyle, J. A., Zavody, A. M., Harries, J. E. and Moffat, P. H. (1983), *Nature* **305** 690–692.

Proffitt, M. H., Faley, D. W., Kelly, K. K. and Tuck, A. F. (1989), *Nature* **342** 233–237.

Quinn, W. H., Neal, V. T. and Antunez, S. E. (1987), *J. Geophys. Res.* **92** 14449–14461.

Ramanathan, V. (1976) *J. Atmos. Sci.* **33** 1330–1346.

Ramanathan, V. (1981), *J. Atmos. Sci.* **38** 918–930.

Ramanathan, V., Cicerone, R. J., Singh, H. B. and Kiehl, J. T. (1985), *J. Geophys. Res.* **90** 5547–5566.

Ramanathan, V., Cess, R. D., Harrison, E. F., Minnis, P., Barkstrom, B. R., Ahmad, E. and Hartmann, D. (1989), *Science* **243** 57–63.

Remsberg, E. E., Russell, J. M., Gille, J. C., Gordley, L. L., Bailey, P. L., Planet, W. G., and Harries, J. E. (1984), *J. Geophys. Res.* **89(D4)** 5161–5178.

Rott, H. and Soegaard, H. (1987), 'Spectral reflectance of snow-covered and snow-free terrain in Western Greenland', *Zeitschr. f. Gletscherkunde und Glacialgeologie*, **23** 115–121.

Russell, J. M., Gille, J. C., Remsberg, E. E., Gordley, L. L., Bailey, P. L., Fischer, H., Girard, A., Drayson, S. R., Evans, W. F. J. and Harries, J. E. (1984a), *J. Geophys. Res.* **89(D4)** 5115–5124.

Russell, J. M., Solomon, S., Gordley, L. L., Remsberg, E. E. and Callis, L. B. (1984b), *J. Geophys. Res.* **89(D5)** 77267–77275.

Russell, P. B., Swissler, T. J. and McCormick, P. (1979), *Appl. Opt.* **18** 3783–3797.

Saltzmann, B. (Ed.) (1983), *Theory of Climate,* Academic Press, USA.

Schiffer, R. A. and Rossow, W. B. (1985), *Bull. Am. Met. Soc.* **66** 1498–1505.

Schneider, S. H. (1987), *Sci. Am., 256* 72–81.

Smagorinsky, J. (1983), 'The beginnings of numerical weather prediction and general circulation modeling: early recollections', in Saltzmann (Ed.), (1983).

Solomon, S. (1988), *Rev. Geophys. 26* 131–148.

Solomon, S., Russell, J. M. and Gordley, L. L. (1986), *J. Geophys. Res.* **91(D5)** 5455–5464.

Stewart, R. H. (1985), *Methods of Satellite Oceanography,* University of California Press, USA.

Stouffer, R. J., Manabe, S. and Bryan, K. (1989), *Nature* **342** 660–662.

Stratospheric Ozone Review Group (1987), *Stratospheric Ozone,* HMSO.

Stratospheric Ozone Review Group (1988), *Stratospheric Ozone 1988,* HMSO.

T-SAT (1988), Report on the Technology Satellite Design Study, Rutherford Appleton Laboratory report RAL-88-033.

Taylor, F. W. (1987), *Surveys in Geophysics* **9** 123–148.

Tucker, C. J. and Miller, L. D. (1977), *Photogramm. Eng. Rem. Sensing of Environment* **43** 721–726.

Tyndall, J. (1861), *Phil. Mag.* **22** 169–194 and 273–285.

Ulaby, F. T., Moore, R. K. and Fung, A. K. (1981), *Microwave Remote Sensing: Active and Passive,* Vol. I, Artech House Inc., USA.

Ulaby, F. T., Moore, R. K. and Fung, A. K. (1982), *Microwave Remote Sensing: Active and Passive,* Vol. II, Artech House Inc., USA.

Ulaby, F. T., Moore, R. K. and Fung, A. K. (1986), *Microwave Remote Sensing: Active and Passive,* Vol. III, Artech House Inc., USA.

United Nations Environment Programme (1989), *Scientific Assessment of Stratospheric Ozone, 1989.*

Untersteiner, N. (1984), 'The Cryosphere', in Houghton (Ed.) (1984).

US Standard Atmosphere (1976), Document NOAA-S/T76-1562, available from US Government Printing Office, Washington DC, 20402.

Vallis, G. K. (1988), *J. Geophys. Res.* **93(C11)** 13979–13991.

Wang, W. C., Weubbles, D. J., Washington, W. M., Isaacs, R. G. and Molnar, G. (1986), *Rev. of Geophys.* **24** 110–114.

Waters, J. W. (1976), 'Absorption and emission by atmospheric gases', in Meeks (Ed.) (1976).

Waters, J. W. (1989), *Atmos. Res.* **23** 391–410.

Wetherald, R. T. and Manabe, S. (1981), *J. Geophys. Res.* **86** 1194–1204.

Wood, J. (1983), *Phil. Trans. Roy. Soc. Lond.* **A309** 337–359.

Woods, J. (1984), 'The upper ocean and air–sea interaction in global climate', in Houghton (1984).

World Climate Programme Report, WCP-95, (1985), 'International satellite cloud climatology project (ISCCP)', World Meteorological Organization, Geneva.

Wunsch, C. (1989), *J. Atmos. Sci. Ocean. Tech.* **6** 891–907.

Wunsch, C. and Gaposchkin, E. M. (1980), *Rev. Geophys. and Space Phys.* **18** 725–745.

Index